Electronic Materials and Application International Conference (EMA 2023)

Orlando, Florida, USA
17-20 January 2023

ISBN: 978-1-7138-7394-5

Printed from e-media with permission by:

Curran Associates, Inc.
57 Morehouse Lane
Red Hook, NY 12571

Some format issues inherent in the e-media version may also appear in this print version.

Printed with permission by Curran Associates, Inc. (2023)

For permission requests, please contact the American Ceramic Society
at the address below.

The American Ceramic Society
550 Polaris Parkway, Suite 510
Westerville, Ohio 43082

Phone: 1-866-721-3322 (toll free U.S. only)
1-614-890-4700 (outside U.S.)
Fax: 1-614-899-6109

ceramics.org

Additional copies of this publication are available from:

Curran Associates, Inc.
57 Morehouse Lane
Red Hook, NY 12571 USA
Phone: 845-758-0400
Fax: 845-758-2634
Email: curran@proceedings.com
Web: www.proceedings.com

TABLE OF CONTENTS

PLENARY SESSION I

Discovery and design of electronic materials using high-throughput experimental methods 1
A. Zakutayev

S2: ADVANCED ELECTRONIC MATERIALS: EFFECTS OF LOW-TEMPERATURE PROCESSING, DOPANTS, AND TEXTURING ON FERROELECTRIC AND PIEZOELECTRIC SYSTEMS

Advancing Solid State Reaction Science Through in situ X-ray Diffraction and Processing Control 2
J. L. Jones, L. Bellcase, C. J. Harper

Dielectric Properties and Redox Behavior of Thick Film and Laminated BNT-based Dielectrics for MLCC Application 3
H. Park, S. Choi, S. A. Khan, S. Lee, J. Lee, W. Jo

High Piezoelectric Properties of [001]-textured (Na, K)(Nb, Sb)-(M, Zr)-(Bi, Ag, Zr) Lead-free Piezoceramics (M = Ba, Sr and Ca) 4
S. Nahm, S. Go, H. Kim, S. Chae, E. Kim

Updates to Cold Sintering of (Na,Ba)(Nb,Ti)O3 5
J. Andrews, R. Boston

Enhanced Piezoelectric Properties of [001]-textured CuO-added PMnN-PZT Ceramics Sintered at Low Temperature.. 6
E. Kim, S. Nahm

Blacklight rapid sintering of ferroelectric ceramics 7
M. Scherer, L. Fulanovic, M. Ameres

Investigating the Grain Growth Kinetics of Sn-Modified PZT 95/5 8
C. M. Culbertson, E. Maclaughlin, A. Dean, D. Zschiesche, T. Baker, C. Roy, B. J. Vitale, P. M. Price, S. Bishop, E. Neuman

S2: ADVANCED ELECTRONIC MATERIALS: DIELECTRIC, FERROELECTRIC, MAGNETIC PROPERTIES AND KINETICALLY STABILIZED MATERIAL SYSTEMS

Synthesis and Properties of Rare Earth Doped Half-Metal Double Perovskite A2FeReO6 (A=La, Pr, Nd) Nanostructures 9
Q. Tang, X. Zhu

Double perovskite oxide Sr2Fe0.5Hf1.5O6-δ: Synthesis, microstructure, dielectric, magnetic and optical properties 10
Q. Tang

Magnetic Bulk Photovoltaic Effect: Strong and Weak Field 11
Z. Dai, A. M. Rappe

Study of dielectric behaviour of La doped Sr2MnO4.. 12
S. Upadhyay

Thickness Dependent Properties of Ferroelectric Boron-Substituted Aluminum Nitride Thin Films 13
J. Hayden, S. Trolier-McKinstry, J. Maria

Impact of Duty Cycle on the Endurance of Hafnium Zirconium Oxide Thin Film Capacitors 14
M. K. Lenox, S. Fields, S. T. Jaszewski, J. Ihlefeld, N. Shukla

Structural characterization of metastable tin(ii) oxide perovskites 15
E. Gabilondo, S. O'Donnell, P. Maggard, R. Newell, R. Broughton, J. L. Jones

Processing and Electrical Characterization of Bulk Nano-Grained Hafnia-based Ceramics 16
B. N. Feigelson, E. Patterson, S. C. Mills, J. Wollmershauser, K. P. Anderson, A. Anguish, J. Andrew

Development of Calcium-doped La(B1, B2, B3)O3 Medium Entropy Perovskite Oxides 17
S. Gajjala, G. Swift, R. Koc

Development of Calcium-doped La(Co0.2Cr0.2Fe0.2Mn0.2Ni0.2)O3 High Entropy Perovskite Oxides 18
S. Gajjala, G. Swift, R. Koc

S3: FERROELECTRICS: GROWTH, STRUCTURE, AND DOMAINS

Unravelling the domain formation during the growth of epitaxial ferroelectrics 19
M. F. Sarott

Strain induced two-step phase transition for ferroelectric hafnia 20
S. Zhou, J. Zhang, A. M. Rappe

Compositional dependence of coercive field, remnant polarization, and breakdown strength in non-equilibrium, magnetron sputtered, ferroelectric zinc magnesium oxide 21
D. K. Goodling, J. Maria

Scanning nitrogen-vacancy microscopy for ferroelectrics and multiferroics 22
W. Huxter, C. Degen, F. Sarott, M. Trassin

Rational design, synthesis, and characterization of organic-inorganic hybrid ferroelectrics 23
W. Xu, K. Romanyuk, P. Zelenovskiy, J. Rocha, A. Kholkin, A. Ushakov, V. Shur

Crafting novel low-symmetry and topological structures in ferroelectric oxide thin films through superlattices 24
D. Sando

Mechanical properties of ferroelectrics and ferroelectric domain walls 25
N. Domingo Marimon

S3: MAGNETISM, MAGNETOELECTRICITY, AND MULTIFERROICS

Multiferroic coupling beyond two ferroic orders 26
M. C. Weber

Ultrafast optically induced perturbation of oxygen octahedral rotations in multiferroic BiFeO3 thin films 27
N. Li, H. Lee, D. Sri Gyan, Y. Ahn, P. G. Evans, E. Landahl, J. Carnis, J. Lee, T. Kim, S. Unithrattil, J. Jo, S. Chun, S. Kim, S. Park, I. Eom, C. Adamo, D. Schlom

Integration of the multiferroic Bi3Fe2Mn2Ox layered supercell on silicon in thin film form 28
J. P. Barnard, H. Dou, Y. Zhang, J. Shen, H. Wang, R. L. Paldi, M. Kalaswad, R. Sarma, A. Siddiqui, Z. He, D. Zheng, P. Ye, N. R. Dilley

Influence of Interfaces and Sub-Unit-Cell Defects on Multiferroic Properties of Aurivillius Phase Films 29
L. Colfer, M. Schmidt, B. Sheehan, L. Keeney, N. Bagués, D. W. McComb, M. Bansal, T. Maity, B. Long

Ultra-Compact Magnetoelectric Nanowire Antennas 30
N. D. Ferson, J. Ganiban, J. Andrew, D. Arnold

Domain specific interactions in hafnia based ferroelectric-ferromagnetic heterostructures 31
B. A. Rao

Elasto-spin modulation of antiferromagnetic ordering in Sr2IrO4 32
J. Liu

Magnetic Structure Evolution in the Entropy-Stabilized Perovskite Oxide (Y0.2La0.2Tb0.2Pr0.2Nd0.2)MnO3 33
N. Arndt, K. Sahebkar, K. Puri, M. Butala, B. L. Musicó, V. Keppens, Q. Zhang, A. Mazza, T. Z. Ward, R. Need

Superconducting epitaxial thin films of (111) KTaO3 grown by hybrid PLD 34
J. Kim

S4: FILM GROWTH AND ADVANCED CHARACTERIZATION

Coherent Imaging and Control of Antiferromagnetic Domain Dynamics 35
C. Ahn

Rotating lattice microstructure in solid-phase epitaxial lateral crystallization of complex oxides 36
R. Liu, D. Sri Gyan, P. Zuo, S. Marks, D. E. Savage, S. E. Babcock, P. G. Evans, T. Zhou, M. Holt, Z. Cai, S. Butun, S. Lu, N. Basit, X. Hu, T. Abbott, N. Kabat, T. F. Kuech

Epitaxial Growth of Hexagonal BaNiO3 Thin Films on SrTiO3(111) Substrates 37
K. P. Koirala, M. Bowden, Y. Du, L. Wang, B. Paudel, T. Kaspar, P. Sushko, S. Chambers, Z. Zhuo, W. Yang

Dynamics of Oxygen Vacancies in SrCoO3-δ Heterostructures 38
D. Fong, Q. Zhang, G. Wan, E. Dufresne, Y. Dong, H. Liu, H. Zhou, A. Sandy, I. Almazan, G. Hu, V. Starchenko, H. Jeen, K. Saritas, J. Krogel, F. Reboredo, H. Lee, P. Ganesh

Unique Co2+ stabilization in oxygen sponge SrFe0.5Co0.5Ox 39
H. Jeen

Quantifying the Oxygen Stoichiometry of a Complex Oxide 40
S. Chambers, M. Hossain, K. P. Koirala, M. Bowden, P. Sushko, Y. Du, L. Wang, W. Samarakoon, H. Zhou, W. Yang

Structural mechanism of unique magnetic properties in patterned complex oxide heterostructures 41
H. Wang, G. Laskin, H. Boschker, V. Srot, J. Mannhart, P. Van Aken

S4: ELECTRONIC AND MAGNETIC PROPERTIES

Strained SrNbO3 becomes a Dirac semimetal 42
H. Lee

Anisotropic Transport at Epitaxial KTaO3 (111) Interfaces 43
A. Al-Tawhid, S. Poage, A. Gonzalez, K. Ahadi

Composition driven tunability of the spin Hall effect in high entropy oxides 44
S. S. Almishal, J. Maria, P. Kezer, D. Mischenko, C. M. Rost, J. Heron

Perovskite Stannate Two-Dimensional Electron Gas with High Mobility at Room Temperature 45
H. Paik, A. Verma, D. Jena, D. Schlom

S4: HIGH ENTROPY OXIDES

Investigation of cation valence in (LaPrCeSmY)0.2O2 high entropy oxide 46
G. R. Bejger, M. Caucci, G. Kotsonis, J. Maria, S. Sinnott, C. M. Rost

Synthesis of Compositionally Complex A6B2O17-form Oxide Thin Films 47
R. Spurling, J. Maria

Electrocaloric Effect of Perovskite High Entropy Oxide Films 48
Y. Son, W. Zhu, S. Trolier-McKinstry

Insights on vacancy formation in entropy stabilized oxides from first principles calculations 49
S. Chae, L. Williams, J. Lee, E. Kioupakis, J. Heron

Correlated Electron Metals with High Entropy as Transparent Conductors 50
S. S. Almishal, F. M. Vieira, S. V. Ayyagari, R. Zu, V. Gopalan, N. Alem, I. Dabo, J. Maria, P. Kezer, J. Heron

Tunability of Lattice Parameter by Means of Valence Diversity in HEOs 51
J. P. Barber, G. Niculescu, D. A. Mischenko, C. M. Rost, M. Webb, J. Heron

S6: WIDE BANDGAP AND ULTRAWIDE BANDGAP SEMICONDUCTOR THIN FILMS AND HETEROJUNCTIONS

Epitaxial Cubic Boron Nitride Grown by Ion Beam-Assisted Molecular-Beam Epitaxy on Diamond 52
D. F. Storm, S. Maximenko, N. Nepal, D. J. Meyer, A. Lang, T. I. Feygelson, B. B. Pate

Thermal transport of (U)WBG semiconductors 53
J. Shi

Diamond Wide Band Gap Material for Quantum Devices and Applications 54
L. Narayanan Ramasubramanian, M. Singh, R. N. Singh

InN and GaN Thin Films via Sputter Deposition 55
J. Nordlander, J. Maria, R. Collazo, Z. Sitar

SiC wafer cost reduction with remote epitaxy technology 56
T. Maekawa

MOCVD growth of Gallium Oxide and Related Materials 57
F. Alema, A. Osinsky

S6: CHALCOGENIDE SEMICONDUCTORS WITH NOVEL OPTICAL AND ELECTRICAL PROPERTIES

Thin film Growth and Quantum Confinement of the Intrinsic Topological Superlattice Bi2-Bi2Se3 58
J. P. Corbett, M. Brown, T. Muratore, R. Laing, K. Burzynski, V. Sreenivas, K. Eyink, A. Reed, K. Mahalingam, D. Winner, J. Jay Gupta, J. Guerrero-Sanchez

Charge and strain tunable magnetism at the MnTe/InP interface 59
M. Chilcote

Probing the electron-phonon dynamics around the band gap of amorphous silicon telluride 60
M. Islam, K. Aryana, D. Hirt, J. Tomko, P. Hopkins, D. Stewart, J. Nag, M. Grobis

3D structure–property correlations of emerging ferroelectric materials revealed by tomographic atomic force microscopy 61
J. Song, B. Huey, S. Wang, L. You, A. Chen, S. T. Jaszewski, J. Ihlefeld

S6: STABLE, ALL-INORGANIC PEROVSKITE SEMICONDUCTORS

Homogeneously Miscible Fullerene inducing Vertical Gradient in Perovskite Thin-Film toward Highly Efficient Solar Cells 62
I. Jeon

Band-gap engineering of ABO3 (A = Ba) perovskites by isovalent B-site substitution 63
M. Braun, S. Wagner, M. Hinterstein, M. J. Hoffmann, Y. Pramudya

Entropy as a control over defects, doping, and phase in strongly correlated materials 64
A. Mazza

Atomically Engineered Thin Films and Devices 65
S. Xie

S7: SUPERCONDUCTING MATERIALS: THEORY, SYNTHESIS AND CHARACTERIZATIONS

Designing High-Tc Superconductors with BCS-inspired Screening, Density Functional Theory and Deep-learning 66
K. Choudhary

Search for Advanced Electric Conductors – an Emerging Challenge 67
T. J. Haugan, M. Sebastian, J. Bulmer, T. Bullard

Monolayer superconductivity and tunable topological electronic structure at the Fe(Te,Se)/Bi2Te3 interface 68
M. Brahlek, Y. Pai, M. Chilcote, H. Miao, S. Okamoto, M. McGuire, R. Moore, X. Yao, S. Oh

The pairing symmetry in quasi-one-dimensional superconductor Rb2Mo3As3 69
D. Arcon, Z. Gosar, T. Arh, A. Zorko, B. Lv

Competing Incommensurate Spin Fluctuations and Magnetic Excitations in Infinite-Layer Nickelate Superconductors 70
C. Lane, R. Zhang, J. Sun, B. Barbiellini, R. S. Markiewicz, A. Bansil, J. Zhu

Non- and near-contact Andreev reflection for quantitative microscopy of superconductors with atomic resolution 71
P. Maksymovych

Toward control and manipulation of superconducting vortex lattices from nano to mesoscales 72
S. Song, J. Yan, E. Dumitrescu, G. Halasz, C. Hua, B. Lawrie, P. Maksymovych, W. Ko

Tunable Electronic States and Instabilities in PbSnTe Heterostructures 73
A. Al-Tawhid, A. Gonzalez, S. Poage, K. Ahadi

S7: MAGNETIC MATERIALS AND HETEROSTRUCTURES

Magnetic imaging of domain walls and surface magnetism in antiferromagnetic topological insulators 74
W. Wu

The electronic detection of quantum spin fluctuations in geometrically frustrated magnet heterostructure 75
H. Zhou

Synthesis and Design of Low-Dimensional Chalcogenides for Infrared Nonlinear Optics........ 76
B. Oxley, A. Iyer, M. Kanatzidis, J. Cho, J. Jang, M. Waters, J. Rondinelli, J. He, V. Gopalan

The field induced quantum spin liquid behavior in a spin-1/2 honeycomb magnet Na2Co2TeO6 77
J. Ma, G. Lin, H. Zhou, Z. Liu

Three-stage magnetic phase transition in ultrahigh-quality 2D CrSBr magnet 78
B. Lv, W. Liu

Transport spin polarization of Pt/ferromagnetic insulator bilayers........ 79
M. Osofsky, J. Prestigiacomo, P. Li, Y. Suzuki

Imaging and Manipulation of Magneto-Acoustic Interactions in Magnetostrictive Devices 80
M. Newburger, M. Page, A. Franson, D. Bas, M. Wolf, P. Shah

S10: DEFECT MOBILITY AND TRANSPORT BEHAVIOR

Probing Ion Migration in ABX3 Perovskite Compounds: Five Fallacies of Simulations 81
R. A. De Souza

Tuning functional properties of complex oxides via defect engineering 82
L. Wang

Effects of Compositional Phase Separation on Mass Transport in Resistive Memory 83
J. Li, W. Sun, Y. Li

Anion intermixing and transport during oxide thin film deposition revealed with isotopic tracers 84
T. Kaspar, S. Taylor, L. Wang, Y. Du, K. Yano, S. R. Spurgeon, D. Schreiber, P. Hatton, B. P. Uberuaga

Tailoring charge transport in mixed conducting oxide thin films via crystallization 85
H. Buckner, J. Simpson-Gomez, N. H. Perry, H. Zhou

Modification of Schottky barrier and electrical properties of oxide ceramics by dislocations 86
M. Soleimany, T. Frömling, L. Porz, E. Bruder, J. Rödel, M. Alexe

Dislocation-tuned electrical conductivity in functional ceramics 87
Q. Muhammad, T. Frömling, J. Rödel

S10: STRUCTURE OF DEFECTS AND DEFECT MEDIATED PROPERTIES

Atomistic Mechanism of Lithium Ion Diffusion at Grain Boundaries and Surfaces in Li-Metal Oxides 88
Y. Ikuhara

Defect-focused, high-throughput screening of ionic & mixed conductors 89
N. H. Perry

Determination of point defects with Dark Field X-ray Microscopy (DF-XRM) 90
A. P. Gayoso Padula, T. Ræder, H. Simons, M. Höfling, R. Rodriguez Lamas

Oxygen vacancy migration and impact on high voltage DC polarization in BaTiO3-Bi(Zn,Ti)O3 91
S. Bishop, M. Blea-Kirby, A. Peretti, L. Jauregui, W. Bachman, J. A. Bock

Fundamental Study of Defect Migration under Burn-in Conditions for High Reliability in Multilayer Ceramic Capacitors 92
C. Randall, P. Yousefian

Determining the Effect of Burn-in Process on Reliability of X7R Multilayer Ceramic Capacitors 93
P. Yousefian, C. Randall

Electrical Characterization of Vanadium and Yttrium Co-Doped BaTiO3 94
A. Aumen, M. Frame, E. C. Dickey

Doping Mechanisms and Electrical Properties of Sodium Niobate 95
J. Canton, D. C. Sinclair

Defect-induced, ferroelectric-like switching and adjustable dielectric tunability in antiferroelectrics 96
H. Pan, Z. Tian, M. Acharya, X. Huang, L. W. Martin

S14: MAGNETIC MATERIALS AND APPLICATIONS

0-3 Magnetic Composites for Power Electronics, Sensing, and Radio-Frequency Applications 97
C. S. Smith

Investigation of optimal pulsed laser deposition parameters for the growth of single crystal barium hexaferrite thin films 98
S. C. Mills, M. Staruch, M. Frye, L. M. Garten

Magnetism in Epitaxial Uranium Dioxide Thin Films 99
A. Chen

Anomaly in Hall resistance: Topological Hall effect or two-channel anomalous Hall effect? 100
P. Roy, Q. Jia, A. Chen, H. Zeng

Voltage-Controlled Topotactic Phase Transformations in Perovskite Cobaltites for Magneto-Ionics 101
C. Leighton

S14: GROWTH AND CHARACTERIZATION OF FERROIC MATERIALS

Magnetic garnets for spintronic and photonic devices 102
C. Ross

Active Tayloring of Funtional Oxides to Couple Electronics to Spintronics... 103
D. A. Gilbert

Polarization dynamics during multiferroic oxide thin film growth 104
M. Trassin

S16: IN SITU/OPERANDO CHARACTERIZATION OF NANOMATERIALS I

Insights on Structure-Property Correlations in Memristive Devices... 105
L. Molina-Luna, R. Winkler, A. Zintler, D. Nasiou, O. Recalde-Benitez, T. Vogel, E. Piros, N. Kaiser, L. Alff

In-situ Characterization of Nanomaterials Using Advanced Electron Microscopy 106
Y. Zhu

Revealing strain evolution in catalytic ceria nanoparticles via in-situ Bragg Coherent Diffraction Imaging.. 107
P. Donahue, C. Carr, A. Popescu, S. M. Haile, T. D. Frazer, K. Harmon, W. Cha, S. Maddali, Y. Cao, S. Hruszkewycz, Y. Liu

High Temperature X-Ray Diffraction of Sputtered Thermite Reactive Nanolaminates 108
C. Skidmore, J. Maria

Surface-Oxide-Driven Evolution of the Anomalous Hall Effect in MnBi2Te4 Thin Films: Implications for the Quantum Anomalous Hall Effect 109
M. Brahlek

New Applications of Direct Detection, Electron Counting Cameras for Materials Science.............................110
J. Moering

S16: IN SITU/OPERANDO CHARACTERIZATION OF NANOMATERIALS II

Operando two-terminal devices inside a transmission electron microscope 111
O. Recalde-Benitez, T. Jiang, R. Winkler, D. Nasiou, Y. Ruan, E. Adabifiroozjaei, A. Zintler, A. Arzumanov, L. Alff, P. Komissinsky, L. Molina-Luna, T. Van Omme, Y. Pivak, H. Perez-Garza, W. Hubbard, B. Regan

In-situ characterization of perovskite oxides with neutron reflectivity 112
A. Mazza

Probing Structure Evolution in Functional Oxides by in-situ Neutron Diffraction 113
Y. Chen, K. An

Thermal Behavior Transition of Graphite under High Pressure 114
Y. Wang, Z. Ye

Ferroelectric Doped Bismuth Ferrite for Photocatalysis 115
A. Dubey, C. Keat, V. V. Shvartsman, M. Escobar, D. C. Lupascu, K. Yusenko, A. Buzanich, U. Hagemann, S. Kovalenko, J. Stähler

Operando measurement on lithium-ion batteries using a laboratory X-ray diffractometer 116
K. Saito, S. Bates

AXON: Experiment-to-Analysis Software for High Performance Transmission Electron Microscopy Studies 117
M. Dressel Dukes, K. Marusak, S. Walden, Y. Guo, J. McConnell

POSTER SESSION

Thermoelectric performance and high-field behavior of Pb-free PTCRs 118
M. Stueckler

Relating the initial microstructural state's grain boundary properties to the abnormal grain growth behavior in SrTiO3 119
V. Muralikrishnan, A. Krause, J. Langhout

Understanding Metal-Insulator Transitions in Heteroanionic TiOF with ordered Heteroleptic Polyhedra 120
S. Nathan, J. Rondinelli

Novel Integration of Graphene-based Ceramics for Enhanced Supercapacitor Energy Storage 121
J. Strimaitis, S. Danquah, C. Denize, S. Pradhan, M. Bahoura

Grain Boundary Chemistry and Potentials in Co-doped BaTiO3 122
S. Hong, E. C. Dickey

Predicting the performance degradation of advanced devices exposed to ionizing radiation 123
X. Guan, M. R. Tonks

Modeling ErAs nanoparticles in III-Vs using Density Functional Tight Binding 124
A. Ram, R. Hu, D. Ho, A. Janotti, G. Bryant

Phase Control of Two-Dimensional Tin Sulfide Compounds Deposited via Atomic Layer Deposition 125
D. Kim, J. Lee, J. Choi, J. Anh

Effects of Y Doping on Ferroelectric and Electrical Properties of As-deposited Hf1-xZrxO2 Thin Films via Atomic Layer Deposition 126
Y. Oh, H. Kim, J. Anh

Rutile-TiO2 thin films deposited by Atomic Layer Deposition using thin SnO2 seed layer and Sn doping for DRAM capacitor 127
M. Jeong, S. Lee, J. Ahn

Processing Optimization of Novel High Tc Piezoelectric Ceramic 128
B. N. Richtik, T. Rowem. Dolgos

The Role of Stoichiometry in Mn1-xZnx Ferrite Microwave Absorbers 129
A. S. Padgett

Flash Sinering of BaTiO3 130
D. Lewin, I. Michiels, K. Menze, D. C. Lupascu

Ferroelectric properties measurements of K1-x NaxNbO3 thin films synthesized through aqueous solution deposition 131
M. Mahmoudvand, A. Z. Mohammed, S. Burns, M. Dolgos

Radio-imaging of electrical discharge in ceramic components 132
J. N. Tilles, P. Clem, A. S. Padgett, R. Martinez

Dielectric properties of co-doped BaTiO3 relaxor ceramics 133
I. Hofbeck, D. Cann

Measurement of Thermal Conductivity of Highly Anisotropic 2D Polymerized Carbon 134
M. Milich, P. Hopkins, E. Meirzadeh, A. M. Evans, C. Nuckolls, X. Roy, B. Foley, A. H. Jones, C. J. Dionne, A. Giri

Applicability of Aerosol Deposition Process in Preparation of Ceramic Coatings 135
B. Xie

Bonding and Crystal Structure Properties of YB66 at High Pressure Conditions 136
R. Kumar, M. Trenary, Y. Xiao, P. Chow, R. Hemley

Conduction mechanism of ZrO2-based nano-laminates structure for suppressing the leakage current 137
S. Lee, M. Jeong, J. Ahn

Piezoelectric Properties of Gadolinium-Modified Bismuth Titanate Ferrite (Bi5Ti3FeO15) for High-Temperature Applications 138
Q. Wang, C. Wang, X. Zhao

Enhanced Energy Storage Properties by Constructing Relaxor Ferroelectrics in Ternary Bi0.5Na0.5TiO3-Bi0.5K0.5TiO3-Ba(Al0.5Ta0.5)O3 Ceramics 139
J. Li, Q. Wang, C. Wang

Frequency dependence of domain switching in lead zirconate titanate 140
R. Beall, R. Broughton, A. Payne, J. L. Jones

Low-Temperature Oriented Bottom Electrodes on Amorphous Surfaces 141
E. C. Fortier, J. Hayden, D. K. Goodling, J. Casamento, J. Maria

In-Situ TEM Experiment of Ferroelectric Switching in AlBN 142
W. Tzou, S. Calderon, J. Hayden, J. Maria, E. C. Dickey

Magnetism in thin films of the high entropy oxide La(Cr0.2Mn0.2Fe0.2Co0.2Ni0.2)O3 143
M. T. Lord, D. Sando, N. Valanoor

Critical Template Characteristics for Efficient and Reproducible Templated Grain Growth (TGG) of Textured Piezoelectric Ceramics 144
E. Suvaci

Author Index

Discovery and design of electronic materials using high-throughput experimental methods

Discovery and design of new materials for electronic applications are at the foundation of technological progress of our modern society. Beyond the classical example of Si-based microelectronics, $Al_xGa_{1-x}As$ enabled optoelectronic devices, $In_xGa_{1-x}N$ revolutionized energy-efficient lighting, and SiC is changing power conversion technologies. Historically, these developments have been slow, laborious, and often serendipitous, taking on average two to three decades from demonstration to market adoption. Two factors are accelerating this rate of progress: automation of synthesis and characterization experiments, and computer aided design of the materials composition and structure. In this presentation, I will overview our recent efforts to discover and design new electronic materials for electronic applications. To this end, we use a combination of thin-film high-throughput (combinatorial) experimental methods and targeted experiments, in tight collaboration with computational and data-driven studies. Potential electronic applications in our focus range from photovoltaic solar cells and photoelectrochemical devices, to neutron detectors and other extreme electronics, to energy-efficient memory and computing. By means of example, I will focus on discovery and design of nitride ferroelectrics with wurtzite crystal structures such as $Al_{1-x}Sc_xN$, for non-volatile memory and high-temperature devices.

Advancing Solid State Reaction Science Through in situ X-ray Diffraction and Processing Control

Our ability to study and advance the science of solid state reactions relies on the tools and techniques with which we can monitor them. However, some basic techniques such as Differential Thermal Analysis and Thermogravimetric Analysis monitor macroscopic phenomena and are insensitive to the microscopics and phase information. In situ XRD during the solid-state reaction is an established method to study the formation mechanisms of perovskites from starting oxides and carbonates. Here, we highlight several examples of fundamental insight garnered from such measurements in important and emerging materials including $Na_{0.5}Bi_{0.5}TiO_3$-$BaTiO_3$, $K_{0.5}Na_{0.5}NbO_3$, (Ba,Sr)TiO_3, and $BiFeO_3$. In NBT, we found that Bi_2O_3 served as a host particle that transforms into the NBT product, a parallel to how TiO_2 converts to perovskite $BaTiO_3$. In KNN, we found that the use of orthorhombic phase Nb_2O_5 starting materials resulted in more chemical homogeneity than monoclinic phase Nb_2O_5. In several other systems, we also show important particle-particle interactions that occur during and influence the formation of the perovskite phase. In summary, we reveal important microstructural interactions during solid state processing that is used to inform process improvements and synthesis consistency.

Dielectric Properties and Redox Behavior of Thick Film and Laminated BNT-based Dielectrics for MLCC Application

Recently, advances in automotive technology such as Hybrid Vehicle(HV), Electric Vehicle(EV) and the electronalization of automobiles have required MLCCs(Multilayer Ceramic Capacitors) with higher breakdown strength and higher working temperature. The $BaTiO_3$-based MLCCs are difficult to satisfy the requirements, so it is necessary to develop a next-generation dielectric for the automotive MLCCs. Therefore, a bulk specimen, single layer chip and multilayer chip were fabricated using a BNT($Bi_{0.5}Na_{0.5}TiO_3$)-based dielectric having a higher phase transition temperature and dielectric constant, and their dielectric properties were investigated. In this study, single-layer-chip and multilayer-chip were prepared by casting green sheets of various thicknesses through the optimization of tape-casting for BNT-based dielectrics. The difference of various dielectric properties according to the shape of the specimen was compared and analyzed. To understand redox behavior, MLCC was sintered in different atmospheres, in which the slurries were optimized by the dielectric powders content and additives' composition. Ag-Pd electrode MLCC was also prepared to check working temperature. Through this study, the possibility of manufacturing MLCC using the BNT-based dielectrics was suggested, and the redox behavior and working temperature of the dielectrics could be confirmed.

High Piezoelectric Properties of [001]-textured (Na, K)(Nb, Sb)-(M, Zr)-(Bi, Ag, Zr) Lead-free Piezoceramics (M = Ba, Sr and Ca)

0.96(Na0.5K0.5)(Nb0.93Sb0.07)O3-(0.04-x)MZrO3-x(Bi0.5Ag0.5)ZrO3 [NKNS-(0.04-x)MZ-xBAZ] piezoceramics (M = Ba, Sr, Ca) are well textured along the [001] direction using 3.0mol% NaNbO3 seeds. The textured-NKNS-0.02MZ-0.02BAZ piezoceramics have a rhombohedral-orthorhombic-tetragonal (R-O-T) structure with a large proportion of the O-R structure (> 80%). These specimens exhibit the large d33 values ranged between 770 pC/N and 805 pC/N, which are the largest d33 values of NKN-based piezoceramics to date. In particular, the textured-NKNS-0.02BaZ-0.02BAZ piezoceramic shows large d33×g33 (29.5×10-12 m2/N) and it exhibits a large strain (0.17% at 4.0 kV/mm). Therefore, this specimen is an outstanding piezoceramic material for piezoelectric energy harvesters (PEHs). A PEH is fabricated using this specimen. The PEH shows a large power density (4.3 mW/cm3), which is the largest value among the PEHs produced by lead-free piezoceramics. Therefore, texturing is an excellent technique for improving the piezoelectricity of NKN-based piezoceramics for PEH.

Updates to Cold Sintering of (Na,Ba)(Nb,Ti)O3

The rapid growth in electronic devices, driven by automation and connectivity in the modern world, requires ~3 trillion ceramic capacitors annually [1]. Barium titanate (BT) based materials are commonly used in multilayer ceramic capacitors, and require a variety of dopants (often rare earth materials) to achieve the X7R industry standard [2][3]. Sodium niobate forms a solid solution with BT of interest for capacitors due to improved temperature stability caused by diffuse phase transitions [4]. High temperatures commonly used to densify ceramics, account for up to 35% of thermal energy input [5]. Cold sintering reduces processing temperature which may lead to benefits in capacitor manufacture, including enabling the use of a wider range of electrode materials and co-sintering of previously incompatible material combinations [6]. In this work we give an update on our cold sintering method which can be used to densify materials in the (Na,Ba)(Nb,Ti)O3 system, aiming to tailor dielectric properties by altering starting powder composition and transient solvent/infill phases.

Enhanced Piezoelectric Properties of [001]-textured CuO-added PMnN-PZT Ceramics Sintered at Low Temperature

The 1.0 mol% CuO-added $0.14Pb(Mn_{1/3}Nb_{2/3})O_3$-$0.86Pb(Zr_{0.55}Ti_{0.45})O_3$ (CPMnN-PZT) piezoceramics were sintered at low temperatures (≤ 950 °C) by the columbite method. It exhibited a high Q_m of 3016 and a high T_C of 309 °C. However, it showed small d_{33} and k_p values of 143 pC/N and 0.43, respectively. The CPMnN-PZT piezoceramic was textured along the [001] direction using $BaTiO_3$ (BT) templates to enhance its soft piezoelectric properties. The [001]-textured CPMnN-PZT + y vol% BT thick films (1.0 ≤ y ≤ 5.0) were densified at a low temperature of 950 °C with a high Lotgering factor (≥ 92 %). The textured sample (y = 1.0) showed a high Q_m of 1400 with increased d_{33} and k_p values of 278 pC/N and 0.58, respectively, and its piezoelectricity showed excellent temperature stability up to 225 °C. Therefore, the textured CPMnN-PZT sample can be used for developing high-power piezoelectric devices.

Blacklight rapid sintering of ferroelectric ceramics

Sintering of ceramics is a thousand-year-known technology. Modern furnaces offer outstanding temperature and atmosphere control for complete densification of ceramics; however, they are energy- and time- inefficient. Recent development of rapid sintering techniques reduced the required energy and time, in particular exemplified by blacklight sintering, featuring the highest energy efficnecy. Ferroelectric ceramics are complex materials whose properties are not just governed by microstructure but also by the delicate defect chemistry. This makes sintering of ferroelectrics even more challenging. To verify the viability of ferroelectric sintering by blacklight, bulk barium titanate ceramics were prepared. This prototype ferroelectric can be sintered in a matter of 15 seconds, with an achieved relative density of 93%. Furthermore, established barium titanate electrical properties are accomplished, as the room temperature permittivity of 5000 is determined with the Curie point at 130 °C. High electric field measurements revealed a ferroelectric hysteresis loop at 20 Hz. Although all samples exhibited a high contribution of leakage current, it was demonstrated that post-annealing minimizes this effect, probably caused by oxygen vacancies.

Investigating the Grain Growth Kinetics of Sn-Modified PZT 95/5

The grain growth kinetics of Sn-modified PZT 95/5 doped with Al were investigated via optical microscopy. Samples were prepared by the traditional solid state synthesis method with up to 3000 ppm aluminum doping, and sintering temperature and time varied from 1150 to 1350 °C and 1 to 45 hours, respectively. Light optical microscopy and digital image analysis was used to characterize the microstructure. Additional characterization methods such as X-ray diffraction, elemental analysis, and several electric property measurements were conducted. The effect of aluminum doping, sintering temperature, and sintering time on the microstructure and grain growth kinetics will be examined.
Sandia National Laboratories is a multimission laboratory managed and operated by National Technology and Engineering Solutions of Sandia, LLC, a wholly owned subsidiary of Honeywell International, Inc., for the U.S. Department of Energy's National Nuclear Security Administration under contract DE-NA000352

Synthesis and Properties of Rare Earth Doped Half-Metal Double Perovskite A2FeReO6 (A=La, Pr, Nd) Nanostructures

In this work, A2FeReO6 oxide (A=La, Pr, Nd) nanoparticles with double perovskite structure were prepared by sol-gel process and solid-state reaction method. The dielectric and magnetic properties of A2FeReO6 oxide (A= La, Pr, Nd) nanostructures with double perovskite structure were characterized. The direct optical band gap of the double perovskite-structured La2FeReO6 (LFRO) oxides was measured to be 2.30 eV, which was also confirmed by their green photoluminescence spectra with a strong peak around 556 nm. Dielectric measurements reveal that LFRO oxides exhibit a relaxation-like dielectric behavior with activation energy (Ea) of 0.05 eV, and the Curie temperature of LFRO is 750 K. And a Griffith-like phase was observed in LFRO powder with a Griffith temperature of 758 K. In addition, we prepare Pr2FeReO6 and Nb2FeReO6 nanoparticles for the first time. Their crystal structures and surface morphologies are characterized as well as the optical, dielectric and magnetic properties. The effect of La substituted by other rare earth elements at A-site on the microstructure and physical properties of the A2FeReO6 oxides (A= Pr, Nd) is investigated and discussed. The present results provide a scientific basis for the design and development of new high-temperature spintronic materials and related devices as well as their applications in photovoltaics.

Double perovskite oxide $Sr_2Fe_{0.5}Hf_{1.5}O_{6-\delta}$: Synthesis, microstructure, dielectric, magnetic and optical properties

In order to develop spintronic devices with high performance, half-metal materials with 100% spin polarizability charge carriers are highly desired, and double perovskite oxides are promising candidates due to their stable ferromagnetism and large spin polarization above room temperature. Many HM DP oxides have been theoretically predicted by first-principles calculations, but only a few of them have been experimentally verified. $Sr_2Fe_{0.5}Hf_{1.5}O_{6-\delta}$ oxide nanoparticles were prepared by solid-phase reaction method for the first time. The Fe/Hf cationic ordering degrees at B-site of SFHO oxides are well controlled by adjusting the annealing temperature and cooling rate of the sample, the differences between B'- and B"-sites in ionic radius and chemical oxide states, which in turn modulates the Curie temperature and spin polarizability. X-ray diffraction and high-resolution scanning transmission electron microscopy were used to reveal the ordering arrangement of B-site ions from inverse space and real space. The influence of B-site cationic ordering degree on the half-metallicity and optical band gap of DP oxides is systematically investigated. Dielectric and magnetic measurement reveal their dielectric behaviors and magnetic properties of the DP oxide samples. These results are useful for developing new high-temperature spintronic materials and photovoltaic devices.

Magnetic Bulk Photovoltaic Effect: Strong and Weak Field

Shift current and ballistic current have been proposed to explain the bulk photovoltaic effect (BPVE), and there have been experiments designed to separate the two mechanisms. These experiments are based on the assumption that under magnetic field, ballistic current can have a Hall effect while the shift current cannot, which is from some unproven energy-scale arguments. A recent work using quantum transport formalism achieves a conclusion that shift current indeed has a Hall current, seemingly contradicting the previous assumption. Moreover, the behavior of BPVE under strong magnetic field is unexplored. In this work, we carry out a systematic numerical study of the BPVE under magnetic field by treating the field non-perturbatively. Our model clearly shows the appearance of the magnetically-induced ballistic current along the transverse direction, and interestingly a sizable longitudinal response of the shift current is also observed, a phenomenon that is not captured by any existing theories. More surprisingly, drastically different shift current is found in the strong-field regime, and the evolution from weak to strong field resembles a phase transition. We hope that our work could resolve the debate over the behavior of BPVE under magnetic field, and the strong-field behavior of shift current is expected to inspire more studies on the relation between nonlinear optics and quantum geometry.

Study of dielectric behaviour of La doped Sr_2MnO_4

The powder and ceramic of the La substituted layered oxides $La_xSr_{2-x}Mn_{1-x}O_4$ (x = 0.3, 0.5, 0.7) were synthesized using the solid-state reaction technique. The Rietveld refinement confirmed tetragonal structure with the space group I4/mmm same as of the Sr_2MnO_4 for all synthesized compositions (x = 0.3, 0.5, 0.7). The SEM images showed clear changes in grain morphology on La substitution. The XPS analysis indicated the presence of Mn in mixed-valence states Mn^{3+}- Mn^{4+}. Studies of permittivity and AC conductivity with respect to temperature and frequency showed negative permittivity over the entire measuring range (30-300 oC). The negative permittivity observed in the 1 kHz-2 MHz frequency range is attributed to plasma oscillations of free carriers, generated by the heat excitation process. A good matching between experimental data (and vs plots) and data generated according to the Drude-Lorentz model confirmed the presence of both localized and free charge carriers in the samples. The inductive character of the samples is shown by the modelling of impedance data using various combinations of resistors and inductors. The inductive character of samples shows the prospect of developing synthesized compositions as coil-less inductor.

Thickness Dependent Properties of Ferroelectric Boron-Substituted Aluminum Nitride Thin Films

Wurtzite structured ferroelectrics are compelling candidates as next generation materials for ferroelectric memory devices, however they suffer from large leakage currents at small film thicknesses. In this work, we investigate the thickness dependent properties of ferroelectric boron-substituted aluminum nitride thin films deposited by magnetron sputtering. Films grown on tungsten coated sapphire substrates show robust ferroelectricity for thicknesses as small as ~100 nm. Polarization hysteresis measurements of films thinner than 100 nm show inflated remanent polarization values from leakage currents. Electron microscopy studies and thickness dependent x-ray rocking curve measurements indicate there is a defect rich region near the film-electrode interface with reduced film crystallinity and out-of-plane mosaicity. In an attempt to improve film quality near the interface, films are grown on a variety of electrode materials with varying chemical and structural similarity to wurtzite structured aluminum nitride. Tungsten nitride is chosen as it is nitrogen containing, yet structurally dissimilar to aluminum nitride, while indium nitride is chosen because it is isostructural to aluminum nitride. Film crystallinity, leakage current, and ferroelectric properties are investigated as a function of indium nitride surface morphology and ferroelectric film thickness.

Impact of Duty Cycle on the Endurance of Hafnium Zirconium Oxide Thin Film Capacitors

The effect of electric field cycling duty cycle on hafnium zirconium oxide (HZO) thin film capacitor endurance is investigated. Capacitors were prepared comprising 10 nm of HZO deposited using plasma-enhanced atomic layer deposition with tungsten top and bottom electrodes. Grazing-incidence X-ray diffraction showed that the ferroelectric orthorhombic Pca21 HZO phase was present. 2 MV/cm field 10 kHz frequency bipolar pulse waves with 4%, 10%, 20%, 40%, and 50% duty cycles were used to cycle devices from a pristine unswitched state. Hysteresis and PUND measurements were taken after each cycling interval. Results showed that increasing the duty cycle decreases capacitor lifetime, with a 4% duty cycle allowing a device to have a lifespan of over 108 cycles while a 50% duty cycles results in only 105 cycles before significant electrical leakage or failure. Increasing duty cycles results in increased wake up behavior at the expense of greater leakage currents. Fatigue prior to failure was not observed for any duty cycle. Earlier failure with respect to duty cycles could be indicative of increased oxygen vacancy formation or accumulation into percolating paths resulting from an increased time under applied fields. This work demonstrates the importance of providing waveform information when reporting endurance data.

Structural characterization of metastable tin(ii) oxide perovskites

Tin has long been pursued as a prospect for replacing lead in prevalent electroceramic materials, however synthesis of such promising tin-containing materials has remained elusive. Recent experiments have enabled synthesis of tin(ii)-containing perovskites, such as $SnHfO_3$[1],but the structural motifs are not yet well-understood beyond bulk characterization. This work describes the structure of metastable tin(ii) oxide perovskites determined via x-ray diffraction, and the local crystal structure determined via transmission electron microscopy. The information will illustrate how the crystal structure of a precursor material determines the structure of the tin-containing reaction product, and more specifically, how the final structure gives rise to an evidently modulated structure, represented by quadrupled unit cell axes (with respect to the cubic perovskite structure). The superlattice reflections are explored in the context of tilt systems and their potential implications on these materials for ferroelectric technologies. Future perspectives are briefly discussed, in particular scaling experiments from nanoscale structural investigations to macroscopic ferroelectric measurements. [1] Gabilondo, E., O'Donnell, S., Broughton, R., Jones, J., & Maggard, P. (2021). Synthesis and stability of Sn (II)-containing perovskites:(Ba, SnII) HfIVO3 versus (Ba, SnII) SnIVO3. Journal of Solid State Chemistry, 302, 122419.

Processing and Electrical Characterization of Bulk Nano-Grained Hafnia-based Ceramics

Thin films of hafnia HfO_2 (or HFO) and hafnia zirconia, $Hf_{0.5}Zr_{0.5}O_2$, (or HZO) with various dopants have been shown to achieve a ferroelectric phase of either metastable polar orthorhombic or rhombohedral. Much work has been accomplished to optimize the thin films polarization and strain hysteresis when below a threshold of 20-30 nm in thickness. However, comparably little work has been done to investigate bulk materials of these phases. In this work, single phase monoclinic, pseudocubic, and mixed-phase nanocrystalline powders of HFO and HZO solid solutions were synthesized via a reverse co-precipitation reaction using an aqueous solution with different hydroxide bases and precursor materials. Next bulk ceramic pellets of 12.5 mm diameter were pressed and densified using an environmentally controlled- pressure assisted sintering (EC-PAS) process. The pure phase monoclinic HFO sample exhibited high density and a crystallite size of only 27 nm according to X-ray diffraction crystallite analysis. The HZO sample exhibited similar grain sizes. The densified pellets' dielectric and ferroelectric properties are measured for these bulk scale nano-grained samples in order to examine how they relate to their thin film counterparts.

Development of Calcium-doped La(B1, B2, B3)O3 Medium Entropy Perovskite Oxides

Calcium-doped La(B1, B2, B3)O3 (Bi = Co, Cr, Fe, Mn, and Ni) Medium Entropy Perovskite Oxide (MEPO) materials are synthesized by the polymerizable precursor method. The impact of variable calcium doping with three B-site transition metals on the unit cell, sintering, and electrical conductivity was studied using X-ray Diffraction (XRD), Energy Dispersive X-Ray Spectroscopy (EDX), and Scanning Electron Microscopy (SEM). Electrical conductivity was measured by the four-probe AC method. The electrical conductivity in all materials is due to p-type small polaron hopping. Electrical conductivity improved with the addition of calcium and various combinations of B-site transition metals.

Development of Calcium-doped La(Co0.2Cr0.2Fe0.2Mn0.2Ni0.2)O3 High Entropy Perovskite Oxides

Calcium-doped La(Co0.2Cr0.2Fe0.2Mn0.2Ni0.2)O3 High Entropy Perovskite Oxide (HEPO) materials were synthesized by the polymerizable precursor method. The effect of calcium doping at various levels on the unit cell, sintering, and electrical conductivity were studied. The electrical conductivity of the present materials is due to a p-type small polaron associated with the B-site transition metals. X-ray Diffraction (XRD) investigated phase purity. Energy Dispersive X-Ray Spectroscopy (EDX) evaluated the uniformity of elements in the materials. A Scanning Electron Microscope (SEM) imaged the sintered microstructure. The four-probe AC method measured electrical conductivity. X-ray Photoelectron Spectroscopy (XPS) analyzed the surface characteristics and valence states of the transition metals.

Unravelling the domain formation during the growth of epitaxial ferroelectrics

Owing to the steady progress in advanced oxide growth techniques, ultrathin ferroelectric oxides can nowadays be grown with a sub-unit cell precision matching that of semiconductors. The appearance of thin-film and interfacial properties in ultrathin oxides that do not arise in their bulk counterpart enable the design of entirely new functional device concepts. In the case of ferroelectrics, their successful integration into nanoscale devices necessitates, however, achieving a deterministic state of the polarization. Realizing this control requires accessing the polarization at the point of its emergence – during epitaxial growth. In our work, we employ nonlinear optics to track the formation and evolution of ferroelectric domains during the epitaxial synthesis of the prototypical tetragonal ferroelectrics $PbTiO_3$ and PZT. In particular, we investigate the roles of the substrate-imposed strain and the electrostatic environment of the growth process on the dynamically evolving polarization, in order to establish targeted domain configurations with reliable ferroelectric properties. Furthermore, by exploiting our unique insights into the domain formation process, we design a distinctive nanoscale domain configuration in PZT films with a composition near the morphotropic phase boundary that shows a multilevel switching behavior of potential interest for neuromorphic concepts.

Strain induced two-step phase transition for ferroelectric hafnia

Hafnia is a promising candidate for next generation ferroelectric devices due to its robust ferroelectricity in even the thinnest film. However, the origin of this robust ferroelectricity is still unknown. Here, we prove that the ferroelectricity arises from a unique two-step phase transition. The first-step antipolar phase transition is induced by strain. When the antipolar amplitude induced in first step-transition is above a threshold, a second-step ferroelectric phase transition will be activated by polar-antipolar coupling. This strong polar-antipolar coupling could offset the depolarization field and thus stabilize the polarization at thinnest film. Our results demonstrate that the strain and polar-antipolar coupling are the origin of robust ferroelectricity in hafnia and provide a new mechanism against depolarization field.

Compositional dependence of coercive field, remnant polarization, and breakdown strength in non-equilibrium, magnetron sputtered, ferroelectric zinc magnesium oxide

Wurtzite ferroelectric materials hold promise for integration with semiconductor processing constraints and for enabling advances in nonvolatile ferroelectric memory. In this study, the metrics of coercive field, remnant polarization, and electrical breakdown strength are quantified as a function of magnesium content in magnetron sputtered zinc magnesium oxide. Ferroelectricity is demonstrated for concentrations exceeding the thermodynamic equilibrium limit for magnesium incorporation in wurtzite zinc oxide. Magnesium is stabilized in this structure at non-equilibrium levels by the high energetics of plasma bombardment. The structure of a high energy state is quenched into film by rapid cooling of the growth interface. In this study, plasma parameters and plasma-to-film working distance are varied to modify growth energetics and degree of magnesium stabilization. X-ray diffraction is used to assess the degree of stabilization of magnesium in this structure. Broad x-ray diffraction peaks, peak shifts, and shoulder peaks indicate the early stages of phase separation. These indicators are valuable in predicting the ferroelectric metrics when viewed in conjunction with magnesium concentration. Low levels of magnesium stabilization and magnesium concentration are associated with low electrical breakdown strength.

Scanning nitrogen-vacancy microscopy for ferroelectrics and multiferroics

The ability to image electric fields at the nanoscale offers unique opportunities in the study of ferroelectric phenomena. With scanning nitrogen-vacancy (NV) microscopy, a technique traditionally used for magnetometry, we measure the electric fields above ferroelectric domains with sub-100 nm spatial resolution. Electrometry is enabled via a gradiometric measurement scheme and quantum sensing protocols that mitigate charge screening at the sensor. We image domain patterns in a prototypical and application relevant thin film ferroelectric (PbZrxTi1-xO3) and a bulk improper ferroelectric (YMnO3) under ambient conditions, without the need of counter electrodes, applied tip-sample voltages, or tip-sample contact. We additionally reconstruct the three-dimensional electric field vector and surface charge density distribution around a domain. With the ability to measure magnetic and electric fields, scanning NV microscopy is an emerging probing technique for noninvasive characterization of ferroelectric and multiferroic materials.

Rational design, synthesis, and characterization of organic-inorganic hybrid ferroelectrics

Organic-inorganic hybrid ferroelectrics have been considered to be promising alternatives to conventional inorganic ferroelectric oxides because they combine attractive features of inorganic and organic materials. Following the development of novel characterization methods and molecular design strategies past decades have witnessed great advances in the discoveries of new hybrid ferroelectrics, Particularly, piezoresponse force microscopy (PFM), which plays an important role in the search for new hybrid ferroelectrics as well as improving their ferroelectric performances relying on domain engineering. Herein, we have designed and characterized several organic-inorganic hybrid ferroelectrics using comprehensive characterization such as DSC, crystallography, dielectric, piezoelectric, and PFM. For instance, we show that high-temperature multiaxial perovskite (Me3NOH)2[KCo(CN)6] (TMC-4) exhibits a large entropy change of 178.79 J×K−1kg−1 at the bond-switching phase transition. Besides, for the first time, we have realized photoisomerization in a solid-state hybrid ferroelectric (Me2NH2)[NaFe(CN)5NO] that allows for the photoswitching between the ferroelectric ground state and the metastable state.

Crafting novel low-symmetry and topological structures in ferroelectric oxide thin films through superlattices

In ferroelectrics, the competition among various degrees of freedom leads to rich phase diagrams as well as the possibility to create topologically protected polarization textures known as ferroelectric solitons. In thin films, the possibility to tune mechanical and electrical boundary conditions by epitaxial strain, substrate orientation, and superlattice design, offers a versatile route to designing such structures. Here, we use these epitaxy techniques to craft new phases and topological textures in the popular multiferroic BiFeO3 (BFO). First, using "anisotropic epitaxy" (i.e., growing epitaxial films on highly miscut substrates) we create a low-symmetry triclinic BFO phase with a 2.4x enhanced piezoelectric coefficient compared with typical rhombohedral-like BFO. Second, by combining BFO with SrTiO3 in carefully designed superlattices, we show evidence for ferroelectric solitons. High-resolution piezoresponse force microscopy and atomic resolution transmission electron microscopy uncover a zoo of topologies, including centre-convergent/divergent topological defects as small as 3 nm. Phase field simulations and effective Hamiltonian computations reveal that some of these structures can be classed as "bimerons" with a topological charge of ±1, while others can give non-integer topological charges, a first observation in a BiFeO3-based system.

Mechanical properties of ferroelectrics and ferroelectric domain walls

The mechanical properties of materials are described by even parity tensors, therefore are believed to be insensitive to space inversion even for non-centrosymmetric materials such as ferroelectrics, i.e., for a ferroelectric material the mechanical response should not depend on whether its polarization is pointing up or down. This situation can change, however, if flexoelectricity (a relevant phenomenon at the nanoscale) is taken into account, because deforming a ferroelectric material in an inhomogeneous way will yield two sources of polarization: the piezoelectric one due to strain, and the flexoelectric one due strain gradients. These two polarizations can be parallel or antiparallel depending on the ferroelectric polarity, which in turn will result in two different electrostatic energy costs of the deformation. At the macroscale, ferroelectric crystals are generally composed by a complex distribution of domains including domain walls. The vigorous research efforts on the functional properties of ferroelectric domain walls has already shown the these almost bidimensional structures possess different functional properties from the host material, including a significant decrease of the young modulus.

Multiferroic coupling beyond two ferroic orders

In multiferroic materials with two ferroic orders, the order parameters and their domain patterns may be rigidly coupled or completely independent. Both of these cases have their merits. With two prototypical examples, we show that in materials with three or more ferroic orders, unusual combinations of coupling and independence are possible. One such material is Dy0.7Tb0.3FeO3. Here, an antiferromagnetic (L) and a ferromagnetic order (M) induce an electric polarization (P) and a trilinear coupling contribution M*L*P to the free energy. We show that this coupling enables the transfer a domain pattern in M to P, and vice versa, while erasing it in the source state. The third order parameter L acts as the "memory buffer" and mediates the transfer. Taking ErMnO3 as second example, we demonstrate that multiferroics with separately emerging magnetic and electric order can have a magnetoelectric coupling even though the magnetoelectric effect is symmetry-forbidden. We reveal that the structural distortions that induce to the ferroelectric polarization also break the balance of the competing superexchange contributions. By doing so the structural distortion links the polar and magnetic orders. In both cases, the multi-order interactions lead to atypical effects on the basis of domain and domain walls. The work demonstrates the significance of exploration in multiferroics beyond the coexistence of ferroelectric and magnetic order.

Ultrafast optically induced perturbation of oxygen octahedral rotations in multiferroic BiFeO3 thin films

The oxygen octahedral rotation (OOR) configuration of the BiFeO3 is important because the OOR pattern modulates the Fe-O-Fe bond angles and thus affects the superexchange interaction between Fe atoms. The modifications of OOR angles in BiFeO3 thin film coupled to the optically induced strain at picosecond timescale were studied by time-resolved x-ray free-electron laser diffraction experiment. The OOR pattern leads to half-order x-ray reflections with reciprocal-space locations and intensities set by the periodicity of the rotation pattern and the magnitude of the rotation, respectively. The intensities of the OOR reflections were measured at a time delay of 16 ps following a femtosecond-duration optical pulse. At this time the acoustic pulse resulting from optical excitation had propagated into the substrate. The intensities were interpreted to determine the change in the OOR angle due to the optical driven expansion. The observed change in intensity was proportional to the optically induced strain, which is consistent with a linear response of the OOR angle to strain. The experimental results were compared with a density functional theory calculation to obtain further insight into the relationship between the OOR and the elastic strain. A simulation of the diffracted intensity as a function of strain using DFT calculated atomic positions agrees well with the measurement.

Integration of the multiferroic $Bi_3Fe_2Mn_2O_x$ layered supercell on silicon in thin film form

The study of thin film multiferroic materials has drawn much attention from the broader research community in recent years due to their innovative applications. A new system, $Bi_3Fe_2Mn_2O_x$ (BFMO), has been the topic of several works showing its magnetic and ferroelectric properties along with a unique single-phase layered supercell structure, obtained by introducing epitaxial strain through substrate selection. Additionally, demands for multiferroic oxides on silicon have driven various works on strain-controlled buffer layer stacks, particularly for practical device applications. In this work, highly epitaxial growth of the BFMO layered supercell on silicon is successfully obtained using a buffer layer stack of TiN/$SrTiO_3$/CeO_2 to effectively tune the strain and overcome the significant lattice parameter mismatch. The multiferroic, microstructure, and optical properties of the BFMO films on silicon are characterized to show preservation of the desired properties and anisotropic behavior. A Ferroelectric Field Effect Transistor (FeFET) is fabricated and probed to demonstrate the practical applications of such a material.

Influence of Interfaces and Sub-Unit-Cell Defects on Multiferroic Properties of Aurivillius Phase Films

Aurivillius phase materials, where ferroelectric perovskite units are interleaved between dielectric $[Bi_2O_2]^{2+}$ layers, are flexible scaffolds for technological applications. One example is $Bi_6Ti_xFe_yMn_zO_{18}$ (B6TFMO), where substitution of magnetic cations into the layered structure creates a rare, room temperature magnetoelectric multiferroic. Previous work has shown interfacing between the $[Bi_2O_2]^{2+}$ and perovskite layers is a key driver to B6TFMO's properties, where magnetic cation partitioning provides pathways for magnetic super exchange. Structural defects such as out-of-phase boundaries additionally augments cation partitioning and leads to unusual polar topologies. This further enhances B6TFMO's potential for memory storage solutions. Here, B6TFMO films were deposited by direct liquid injection chemical vapour deposition on varied sapphire substrates (surface miscut: 0.2o-10o and patterned) to encourage structural defects. X-ray diffraction provides evidence for an increased presence of structural defects within these films compared to films on flat sapphire. Electron energy loss spectroscopy probed the material's local electronic structure, where changes were observed moving through the perovskite layers. Multiferroic properties, examined through piezoresponse force microscopy and superconducting quantum interference device magnetometry measurements, are discussed.

Ultra-Compact Magnetoelectric Nanowire Antennas

The operational frequency of a conventional conductor-based electromagnetic antenna dictates its physical size, generally the antennas size needs to be at least 1/10th the electromagnetic wavelength. This size-dependency inhibits the miniaturization and system integration of antennas used in relatively low frequency applications (<300 MHz, where wavelength > 1 m). To overcome this limitation, we are developing ultra-compact magnetoelectric (ME) nanowire antennas. By implementing ME materials, which couple both ferromagnetic and ferroelectric behavior, the antenna functions via a strain-based electromechanical effect as opposed to a purely electromagnetic effect. This eliminates the traditional size restrictions of charge-based antennas. Here, we have developed novel ME core@shell cobalt ferrite@barium titanate (CFO@BTO) nanowires through a two-step template-assisted process. Magnetic and ferroelectric properties of the nanowires will be presented along with ME coefficients with varying volume fractions of CFO:BTO, showing peak coefficients of 1 Vcm-1Oe-1 under zero bias magnetic field conditions. Furthermore, the ME antenna's efficiency, in terms of the reflection coefficient will be presented. Ultimately, this work presents a new antenna architecture that has great potential for use in telecommunications, IoT sensors, and non-invasive neural stimulation/recording.

Domain specific interactions in hafnia based ferroelectric-ferromagnetic heterostructures

Hafnia based ferroelectric thin-films have recently gained large attention due to their ability to maintain ferroelectricity even at very low film-thickness. However, due to the difficulty in stabilizing their ferroelectric phase, very few researchers have investigated their interaction with magnetic films. These kind of heterostructures are attractive for their magnetic-ferroelectric (multiferroic) coupling, wherein magnetic domains can be efficiently controlled by application of electric field. In this study, we investigate the domain structure of La-doped HfO2 and Hf1/2Zr1/2O2 films grown on an epitaxial La2/3Sr1/3O3 film using piezoresponse force microscopy. The strain exerted on the hafnia films is found to have a strong correlation with the domain structure of hafnia. In addition, we have also investigated the effect of external magnetic field on the ferroelectric domain structure of hafnia. The results of piezoresponse force microscopy and their correlation with structural, electrical and magnetic properties is discussed in detail.

Elasto-spin modulation of antiferromagnetic ordering in Sr2IrO4

Sr2IrO4 is known to show remarkable similarities to high-Tc cuprates, such the quasi-2D antiferromagnetic Mott insulating state, despite their big difference in spin-orbital coupling (SOC). This emergent analogy is partly due to the lifting of the t2g orbital degeneracy by SOC instead of Jahn-Teller effect. On the other hand, the resulted pseudospin-half state has a strongly spin-orbit-entangled wavefunction, which could lead to novel collective phenomena beyond the cuprates. In this talk, I will discuss our recent work that utilized the elasto-x-ray scattering to perform resonant magnetic diffraction on Sr2IrO4 under in situ strain control in addition to temperature and magnetic field. Giant responses of the metamagnetic transition were observed due to the highly effective detwinning of the antiferromagnetic domains by the uniaxial strain. It enables efficient switching of the metamagnetic transition. In contrast, when we apply the strain in a shear configuration to the ab-plane, a new spin modulation that breaks the c-axis translation symmetry emerges with a domain wall-like layer in between the original ones. Our analysis indicates the underlying interaction is an unusual fourth-order anisotropy that is only activated under strain. Our results showcase the rich physics of 5d complex oxides and demonstrate the power of elasto-x-ray scattering in revealing their emergent phenomena.

Magnetic Structure Evolution in the Entropy-Stabilized Perovskite Oxide ($Y_{0.2}La_{0.2}Tb_{0.2}Pr_{0.2}Nd_{0.2}$)$MnO_3$

The configurational disorder in entropy-stabilized (ES) single-phase materials can lead to functional properties that deviate from traditional alloy mixing rules. ES can be applied to specific sublattices of transition metal oxide structures, adding to the rich composition-structure-property phase space. Work on ES oxides (ESOs) has shown the formation of colossal dielectric constants, enhanced magnetic exchange couplings, and mixed phase magnetic structures. We are focused on the latter; combining bulk magnetometry and temperature-dependent neutron diffraction to better understand the magnetic structure formed in an ESO perovskite. Emphasizing on how long-range magnetic order forms despite large disorder of a cation sublattice in ($Y_{0.2}La_{0.2}Tb_{0.2}Pr_{0.2}Nd_{0.2}$)$MnO_3$, (5A)$MnO_3$. Magnetometry data shows two sharp phase transitions and the presence of exchange bias at low temperature, suggesting mixed phase magnetic ground state. Neutron diffraction shows long-range antiferromagnetic magnetic ordering. Diffraction data reveals that the two phase transitions can be attributed to ordering on the Mn and A-site sublattices; the dominant low-temperature magnetic ordering is Pn'ma'. We compare the similarities between the magnetic structure evolution in (5A)MnO_3 to $NdMnO_3$ highlighting the insensitivity of magnetic structures to disorder on the A-site of the perovskite lattice.

Superconducting epitaxial thin films of (111) $KTaO_3$ grown by hybrid PLD

Modern quantum materials are inherently sensitive to point defects, and require a new synthesis route to produce epitaxial oxide thin films and interfaces clean enough to probe fundamental quantum phenomena. The major challenge of research on $KTaO_3$ system is that it is difficult to grow high-quality $KTaO_3$ epitaxial thin films due to high volatility of potassium. Recently, we have developed the hybrid PLD method for electronic grade $KTaO_3$ thin film growth, which successfully achieves this by taking advantage of the unique capabilities of PLD to instantly evaporate Ta_2O_5 in a controlled manner and evaporation of K_2O to maintain sufficient overpressure of volatile species. We successfully synthesized heteroepitaxial $KTaO_3$ thin films on (111)-oriented $KTaO_3$ bulk single crystal substrates with a $SmScO_3$ template by hybrid PLD, followed by a $LaAlO_3$ overlayer. AFM and TEM data show that the $KTaO_3$ thin films are epitaxial to the bulk single crystal $KTaO_3$ (111) substrates with atomically flat surfaces. Electrical transport data show a superconducting transition temperature of ~ 1.5 K. We anticipate that the ability to synthesize high-quality epitaxial complex oxides such as $KTaO_3$ that contain volatile elements will provide a platform for exploring new physics and technological applications arising from unique characteristics such as large spin-orbit coupling.

Coherent Imaging and Control of Antiferromagnetic Domain Dynamics

In this work we examine the dynamics and energetics of antiferromagnetic (AF) domains. A key enabler to understand and control AF domains involves the ability to image their structure and dynamics. The AF order parameter can be imaged in thin films using resonant soft x-ray diffraction, and information about their dynamics can be accessed by x-ray photon correlation spectroscopy (XPCS) measurements using coherent soft x-rays. Recent studies have revealed that the magnetic and electronic phase transitions in NdNiO3 (NNO) can be tuned by confining the NNO layers to reduced dimensions in heterostructures. We image the AF ground state in reciprocal space for atomically layered (NdNiO3)m-(NdAlO3)n heterostructures and show that dimensional confinement enhances the phase fluctuations at AF domain boundaries. The speckle patterns acquired from coherent x-ray scattering contain information on long- and short-range correlations and the 1D nature of AF boundaries in dimensionally confined nickelate layers. As the AF domain boundary approaches the 1D limit, the dynamics are enhanced. This study demonstrates an approach to characterize dimensional effects on long-range order and the ability to control AF domain configurations in oxide heterostructures.

Rotating lattice microstructure in solid-phase epitaxial lateral crystallization of complex oxides

The lateral epitaxial overgrowth of the amorphous form of the model perovskite oxide SrTiO3 can be seeded by selected areas of a SrTiO3 single-crystal substrate, resulting in the formation of a rotating lattice microstructure. Crystallization was probed in lithographically fabricated crystallization environments defined using a patterned layer of Si3N4. Laterally crystallized SrTiO3 exhibits a rotating-lattice microstructure in which the crystallographic orientation evolves continuously as a function of the lateral crystallization distance across the Si3N4 mask. An elastic argument suggests that stress induced by the difference in density between amorphous and crystalline SrTiO3 leads to the nucleation and motion of dislocations producing the rotating lattice microstructure. The rotation rate during crystallization at 450 °C is 25° per µm, and is doubled for crystallization at 550 °C. Potential origins of the greater rotation rate at 550 °C include a temperature dependence of the stress at the amorphous-crystalline interface or an increase in the nucleation or mobility of dislocations at higher temperature. The development of the rotating lattice microstructure has the potential to enable the creation of complex oxides in stress and orientation configurations not available through other epitaxial growth methods and to expand the range of application of complex oxide epitaxy.

Epitaxial Growth of Hexagonal BaNiO3 Thin Films on SrTiO3(111) Substrates

The dynamic tunability of Ni valance states in complex nickelate thin films has led to emergent properties including superconductivity, enhanced electrocatalytic activity, and metal-insulator transition. It remains controversial whether the Ni4+ state can be experimentally stabilized in perovskite structures rather than its bulk hexagonal phase to integrate new functionalities, and high-quality spectroscopy data for Ni4+ are still lacking. Here, by judiciously choosing perovskite SrTiO3(111) as substrates, we examine the growth of BaNiO3 by oxygen-plasma-assisted molecular beam epitaxy. Our X-ray diffraction and high-resolution scanning transmission electron microscopy studies reveal that epitaxial, hexagonal BNO(h-BNO) can be grown under carefully selected conditions, but not with a seemingly hexagonal on the hexagonal interface. Instead, the epitaxial relationship is proven to be h-BNO(11-20)//STO(111) and h-BNO[0001]//STO[-1-12] as a smaller lattice match can be achieved this way. Density functional theory calculations also corroborate that the observed epitaxial relationship leads to the most energetically favorable configurations. Moreover, high-quality X-ray photoelectron and X-ray absorption spectroscopy data display clear differences from Ni3+ and Ni2+ reference spectra. This work provides insights into stabilizing and integrating Ni4+ in complex nickelates to achieve designed functionalities.

Dynamics of Oxygen Vacancies in $SrCoO_{3-\delta}$ Heterostructures

Strontium cobaltite or $SrCoO_{3-\delta}$ is an end member of $La_{1-x}Sr_xCoO_{3-\delta}$, a fuel cell material and oxide catalyst. While the redox behavior of such oxides has been the subject of considerable interest, much concerning the kinetics and dynamics of these materials remain unknown. Utilizing in situ coherent X-ray scattering at the Advanced Photon Source, we monitored speckle from epitaxial $SrCoO_{3-\delta}$ thin films grown on both $SrTiO_3(001)$ and $(LaAlO_3)_{0.3}(Sr_2TaAlO_6)_{0.7}(001)$ as oxygen was incorporated and evolved (switching the environment from O_2 to N_2) to gain insight into the dynamics of oxygen-induced phase evolution in complex oxide materials. We found that the kinetics of the brownmillerite to the perovskite phase transition could be varied from tens of minutes to several hours over a small temperature range (300°C to 350°C), observing pronounced differences between the oxidation and reduction behaviors, the latter involving substantial incubation times to re-nucleate the brownmillerite phase. From X-ray photon correlation spectroscopy performed at the brownmillerite superlattice reflection, we find that the two-time correlation function differs greatly between the two different substrates. We will discuss the kinetics and dynamics of the oxygen ion / vacancy-ordering phase transition and the methods used to distinguish the different atomic and electronic mechanisms taking place.

Unique Co^{2+} stabilization in oxygen sponge $SrFe_{0.5}Co_{0.5}O$

XComplex oxides have gotten lots of attentions due to diverse electronic/magnetic phases and their tunability by external stimuli such as strain, magnetic fields, electric fields, etc. In addition, mixed ionic and electronic conducting behaviors in complex oxides have been applied for many energy devices such as solid oxide fuel cells (SOFCs) and electrochemical sensors, where redox reactions and catalytic activity at the interfaces of gas-solid and solid-liquid play critical roles for the performance. The primary purpose of this presentation is to address surface ion-exchange behaviors that can be used for controlling phase in epitaxial complex oxides. As a model system, we chose oxygen sponge $SrFe_{0.5}Co_{0.5}O_x$ grown by pulsed laser deposition. Upon facile gas reaction at moderate temperature (~400oC), selective reduction in $SrFe_{0.5}Co_{0.5}O_{2.5}$ occurs. This results in stabilization of unique Co^{2+} state largely due to the loss of oxygen in the lattices. Such valence state transition induces structural phase transition and band gap modification. In addition, using real-time x-ray diffraction and dc transport measurements at the moderate temperature, we probe reversible redox reaction in $SrFe_{0.5}Co_{0.5}O_x$, which can be applicable to not only a cathode in SOFCs but also electrochemical sensors due to realization of electronically three distinct states.

Quantifying the Oxygen Stoichiometry of a Complex Oxide

This study illustrates a strategy to determine the oxygen vacancy concentration in a complex perovskite oxide. Our model system – $SrFe_{0.5}Cr_{0.5}O_{3-\delta}$ (SFCO) – was epitaxially grown on $SrTiO_3(001)$ substrate with a plasma assisted molecular beam epitaxy method. Synchrotron based coherent Bragg rod analysis (COBRA) analysis, with sub-Angstrom length scale resolution, demonstrates gradual expansion of the unit cell volume with increasing distance from the SFCO/STO interface. Our analysis suggests that the continuous lattice expansion primarily originates from the gradation of the oxygen vacancy content in the crystal which also led to variable formal charge states for the cations. The Density Functional Theory (DFT) based simulations elucidate the relationship between the vacancy concentration and the magnitude of the expansion of the SFCO lattice. Based on these experimental and theoretical findings, we developed an analytical model to accurately quantify the δ value in $SrFe_{0.5}Cr_{0.5}O_{3-\delta}$. The presented strategy is also applicable for other complex oxides.

Structural mechanism of unique magnetic properties in patterned complex oxide heterostructures

Complex oxide heterostructures have a wealth of magnetic states stemming from specific coupling between spin, charge, orbital, and lattice degrees of freedom. Patterned complex oxide heterostructures have altered lattice structures and reconstructed electronic systems, offering a playground for feasibly engineering magnetic states. Here, we investigate the atomic structure and magnetic behavior of 5 × 5 mm2 sized SrRuO3 (SRO) nanodots array patterned with electron-beam lithography. The SRO nanodots' size is down to 15 nm, nearly an order of magnitude smaller than most previously reported work on perovskite oxides. We find that SRO shows ferromagnetism even in the smallest dots. The Curie temperature and magnetic anisotropy of SRO nanodots arrays depend on the size of SRO dots. In order to study the atomic-scale structural mechanism behind magnetic behaviors using the transmission electron microscope (TEM), an optimized TEM sample preparation method has been developed for preparing high-quality TEM specimens of tiny SRO nanodots. Scanning TEM investigations indicate that structural relaxation and a resulting oxygen octahedral distortion in the patterned SRO layer lead to the nanodots' size-dependent magnetic properties. Furthermore, we reveal a competing mechanism between the deformation and the oxygen octahedral rotation in magnetic behavior.

Strained SrNbO3 becomes a Dirac semimetal

Quantum materials (QMs) with strong correlation and nontrivial topology are indispensable to next-generation information and computing technologies. The exploitation of topological band structure is an ideal starting point to realize correlated topological QMs. Among many classes of QMs, complex oxides offer a rich playground to exploit both correlations of Coulombic interactions between electrons often found in 3d transition metal oxides (TMOs) and SOC from heavy metal-based 5d TMOs, leading to many intriguing physical properties. Hybrid materials with proper balancing of the correlation and SOC offer unique opportunities to discover new properties, and heterostructuring or combining dissimilar materials is a great approach to harnessing the complex interplay between quantum wavefunctions and various control parameters, such as proximity, dimensionality, topology, and symmetry. Here, we report that strain-induced symmetry modification in correlated oxide SrNbO3 thin films creates an emerging topological band structure. Dirac electrons in strained SrNbO3 films reveal ultrahigh mobility (mmax ≈ 100,000 cm2/Vs), exceptionally small effective mass (m* ~ 0.04me), and nonzero Berry phase. Strained SrNbO3 films reach the extreme quantum limit (XQL), exhibiting a sign of fractional occupation of Landau levels and giant mass enhancement.

Anisotropic Transport at Epitaxial KTaO3 (111) Interfaces

An anisotropic 2D superconductivity was recently discovered at KTaO3 interfaces. The superconducting order parameter is sensitive to the interface orientation with critical superconducting temperatures of 0.05 K, 1 K, and 2 K for (100), (110), and (111) interfaces. Furthermore, high quality KTaO3 (111) interfaces show an anisotropic superconducting transition along [110] and [112] directions. The spin splitting and magnitude of spin-orbit coupling is highly sensitive to the crystallographic direction. We grew high quality metallic KTaO3 interfaces, using an oxide molecular beam epitaxy (MBE). The cross-section high-angle annular dark-field (HAADF) imaging in scanning transmission electron microscopy (STEM) shows an atomically abrupt interface wihtout any extended defects. The sheet resistance v.s. temperature demosntrates a a metallic behavior, dR/dT >0, extending to 2 K. The Hall carrier density is ~6×1013 cm−2 at 2 K, resolving a carrier mobility of ~200 cm2/v.s. We observe an anisotropic superconductivity emerges below 1.5 K. The in-plane critical field (HC2) exceeds the Pauli limit and is sensitive to bias and magnetic field directions. Figure 1 shows the in-plane HC2 along [110] and [112] bias directions. We observe anisotropic behavior in both critical temperature and critical field at the KTaO3 interfaces. Interesringly, the HC2 depends on both bias and applied field directions.

Composition driven tunability of the spin Hall effect in high entropy oxides

The spin Hall effect (SHE) is a material property where a charge current produces a transverse spin current. Converting charge to spin opens new possibilities for logic and memory devices employing spin. Low conversion rates are the primary challenge to enabling new technologies. Here we systematically study the evolution of the SHE in novel high-entropy oxides (HEOs) with material composition to elucidate fundamental mechanisms and explore the possibility of synergistically leveraging varying mechanisms to achieve new paradigms in charge to spin conversion efficiency. HEOs enhance the solubility of elements to enable larg chemical tuning and interplay with local disorder. We explore the tuning of Mott scattering (MS), spin-orbit coupling (SOC), and degeneracy breaking through the tolerance factor in epitaxially grown Sr(CrNbMoW)O3 thin films with superb crystal quality. Second harmonic Hall measurements on HEO/Py stacks to quantify the spin Hall angle, via spin-transfer-torques. Preliminary measurements indicate large y-polarized damping-like torque and have a strong correlation with current density. θSH shows an inverse correlation with Cr molar percentage, suggesting MS from Cr is insignificant relative to the intrinsic SOC in this composition. Our results reveal that large changes in SHE can occur with moderate compositional changes for the design of function for a given application space.

Perovskite Stannate Two-Dimensional Electron Gas with High Mobility at Room Temperature

Two-dimensional electron gases (2DEGs) with high mobility at room-temperature (RT) in wide-band-gap oxides are fascinating phenomena for oxide electronics, particularly for the high-electron-mobility-transistors (HEMT) with transparency. The experimental realization of 2DEGs via polarization/modulation doping in oxide heterostructure, however, has far been limited to date. Here, we report the structure and transport properties of 2DEGs in modulation-doped SrSnO3/BaSnO3 heterostructure grown by molecular-beam epitaxy. Utilizing the band diagram based on the 1D-Poisson's equation and the energy band offset between BaSnO3 and SrSnO3, we experimentally achieved the transfer of mobile electrons carrier from the 1.6 nm thick La-doped SrSnO3 to the undoped BaSnO3. The non-destructive capacitance-voltage measurement by the mercury probe at RT indicates the 2D carrier confinement at the SrSnO3/BaSnO3 interface. The RT 2DEG mobility is measured as high as 175 cm2•V−1•s−1 with a 2D carrier density of 1.0×1013 cm-2, which is the highest RT 2DEGs mobility among the modulation-doped perovskite oxide interface. Our work suggests the tremendous potential of stannate 2DEGs for the next generation of oxide nanoelectronics

Investigation of cation valence in (LaPrCeSmY)0.2O2 high entropy oxide

The phase complexities of rare-earth high entropy oxide (YLaCePrSm)0.2 O2 (F1) has been the subject of recent investigation. Bulk F1 samples exhibit both bixbyite and fluroite-type ordering, and the general similarities between these crystal structures adds to the difficulty of characterization in this high entropy system. Ideal bixbyite is of the form A2O3 where 3+ cations are 6 coordinated to oxygen, while fluorite, of form AO2, maintains 4+ cations that are 8 coordinated to oxygen. Prediction from density functional theory (DFT) suggests mixed valence in Pr and Ce, while other elements exhibit a 3+ charge state. In this work, we compare results from DFT to experiment by measuring the L2 and L3 absorption edges of cations using x-ray absorption near edge structure (XANES). Linear combination fitting (LCF) of our samples using custom-doped standards will be used to provide a weighted average of valence states and cation site occupancy.

Synthesis of Compositionally Complex A6B2O17-form Oxide Thin Films

The demonstration of ferroelectricity in SiO2-doped HfO2 has prompted a surge of research into electronic properties of thin film materials. The metastable orthorhombic phase, with uniquely coordinated Hf cation sites, gives rise to ferroelectricity in HfO2, underscoring the importance of stabilizing this structure. Interestingly, the A6B2O17 (A = Zr, Hf; B = Nb, Ta) family of phases also demonstrates a polar orthorhombic Ima2 structure, however we have shown that this is a highly stable equilibrium structure at high temperatures (>1000 C) and can accommodate compositional complexity (i.e. ≥4 cation species). Here, we report on the synthesis of A6B2O17-form oxide thin films through a sputtering-based approach from bulk ceramic targets. The focus of this study is exploring the possibility of ferroelectricity and controlling growth parameters in A6B2O17 thin films. Particular attention will be paid to assessing the crystalline quality of these thin films and beginning to formulate structure-property relationships within this system.

Electrocaloric Effect of Perovskite High Entropy Oxide Films

This paper describes two perovskite high entropy oxide (PHEO) compositions: Pb(Hf0.2Zr0.2Ti0.2Nb0.2Mn0.2)O3 (Mn PHEO) and Pb(Hf0.2Zr0.2Ti0.2Nb0.2Al0.2)O3 (Al PHEO). Powders were prepared by conventional solid state sintering by first pre-reacting the B-site oxides, then adding PbO. Phase pure Mn PHEO powder was obtained following calcination of the mixed powders at 750°C for 240 min; however, secondary phases persisted in PHEO for heat treatments from 750°C to 1200°C. The PHEO undergoes an entropy-driven phase transformation. Thin films of these compounds were synthesized by pulsed laser deposition (PLD) on a lead zirconate titanate seed layer on Pt-coated SiO2/Si. The dielectric response of the Mn PHEO films showed some contribution from space charge polarizability; in contrast, the Al PHEO films showed a slim ferroelectric hysteresis loop and relaxor-like characteristics. Al PHEO had a dielectric permittivity of ~2000 with a loss tangent < 0.05 from 100 Hz to 100 kHz; it had a dielectric maximum at 105±0.5°C and a Burns' temperature of 234±0.5°C. Indirect measurements based on the Maxwell-relations yielded a maximum electrocaloric temperature change of 8.4 K at 180 under the applied electric field of 1186 kV/cm.

Insights on vacancy formation in entropy stabilized oxides from first principles calculations

We combine high-throughput atomistic calculations and linear regression algorithms to investigate the role of local configurational and structural disorder on the thermodynamics of vacancy formation in (MgCoNiCuZn)O-based entropy-stabilized oxides (ESOs) and their influence on the electrical properties. We find that the cation-vacancy formation energies decrease with increasing local tensile strain caused by the deviation of the bond lengths in ESOs from the equilibrium bond length in the binary oxides. The oxygen-vacancy formation strongly depends on structural distortions associated with the local configuration of chemical species. Vacancies in ESOs exhibit deep thermodynamic transition levels that inhibit electrical conduction. By applying the charge-neutrality condition, we determine that the equilibrium concentrations of both oxygen and cation vacancies increase with increasing Cu mole fraction. Our results demonstrate that tuning the local chemistry and associated structural distortions by varying alloy composition acts an engineering principle that enables controlled defect formation in multi-component alloys.

Correlated Electron Metals with High Entropy as Transparent Conductors

Novel conducting complex oxides with high electron correlation are of keen interest for UV-transparent applications, and as hosts for fundamental studies of correlated electron systems. Sr based cubic perovskite oxides, including SrxNbO3 and SrxMoO3, are appealing to explore due to their heavy cation solubility and tunable electrical conductivity and transparency. Utilizing a high configurational entropy formulation, the high symmetry perovskite phase can be stabilized with five cations on the perovskite B-site. This approach leverages local disorder to enhance optical transparency by suppressing interband transitions through state occupancy engineering and enhance conductivity by populating t2g d-orbitals with conduction carriers. Our results show that Srx(Ti,Nb,Cr,Mo,W)O3 thin films grown by pulsed laser deposition are stabilized as single-phase perovskite. Derivative systems with four B-cations instead of five, exhibit chemical and/or phase segregation. Contributions from individual B-cations and their interactions result in unique overall transport properties. For instance, Cr markedly influences crystallinity and electrical transport. A-site vacancies, too, influence electronic and optical properties, and structure stability. The experimental findings are supported by first principles calculations that highlight the effect of the coexistence of cations on the band structure and orbital filling.

Tunability of Lattice Parameter by Means of Valence Diversity in HEOs

Entropy-stabilized oxides (ESOs) are a unique class of materials that bypass the influence of enthalpy in typical phase formation through a drastic increase of configurational entropy. Although most of the current literature focuses on the application prospects of these ceramics and their potential use in a wide range of technologies, we focus on the fundamental characteristics that ultimately govern their behavior. This project examines correspondence between the sample growth conditions and the resulting charge variance and local structure in seven-component ESO thin films. (MgNiCoCuZn)0.167(CrMn)0.0835O thin films were grown on MgO substrates, using pulsed laser deposition, under varying pressure, temperature, and gaseous atmosphere. A second sample series sees the 6th and 7th components vary composition (CrxMn1-x), with x = 0, 0.25, 0.5, 0.75, 1. Using x-ray absorption fine structure spectroscopy, trends are revealed regarding the relationship between local and global structure characteristics.

Epitaxial Cubic Boron Nitride Grown by Ion Beam-Assisted Molecular-Beam Epitaxy on Diamond

Cubic boron nitride possesses an ultra-wide band gap (6.4 eV, indirect); a high thermal conductivity, second only to diamond; and the capacity to be doped both n- and p-type, making it an attractive candidate for high power and high temperature electronic applications. Further, its structural and electronic similarity to diamond holds the potential for novel electronic devices based on c-BN/diamond heterostructures. However, there are numerous challenges associated with growing device-quality layers of c-BN, including the existence of multiple phases of BN; the metastability of the cubic phase at pressures and temperatures typical of vapor-phase growth; and the absence of large-area bulk c-BN crystals, which necessitates growth on non-native substrates. Cubic boron nitride films were grown on (100) oriented IIa diamond substrates by ion beam-assisted molecular-beam epitaxy (MBE). Fourier transform infrared spectroscopy indicates these films are fully cubic. Transmission electron microscopy confirms the presence of an epitaxial c-BN film with the presence of isolated misfit dislocations and no indication of h-BN. The interface between the c-BN layer and the diamond substrate is structurally abrupt, and no interlayer between the c-BN film and diamond substrate is seen. It was found that trace amounts of impurities, such has Mg, facilitate the growth of c-BN on diamond by ion-assisted MBE.

Thermal transport of (U)WBG semiconductors

With wider bandgap and higher breakdown voltage, wide-bandgap (WBG) and ultrawide-bandgap (UWBG) devices are smaller, lighter, faster, more reliable, and hold the potential to revolutionize the electronics industry and pave the way towards an energy-efficient world. However, the reliability and performance of WBG and UWBG electronics are limited by the intense heating, and the thermal conductivities of some WBG and UWBG materials are orders of magnitude lower than that of traditional semiconductors. Therefore, WBG and UWBG devices inevitably confront a much more severe thermal problem. For efficient thermal management of WBG and UWBG devices for futurre power and RF applications, both of the heat transfer in materials and across interfaces of WBG and UWBG electronics are explored with modeling and experimental methods from atomistic-scale to device-level. Different interfaces like semiconductor-dielectric and semiconductor-metal interfaces are studied, and the effects of different mechanisms like growth methods, crystal orientations, and interfacial chemical reactions on energy transport will be discussed.

Diamond Wide Band Gap Material for Quantum Devices and Applications

Diamond is a fascinating material due to its wide band gap, optical transparency, and high thermal conductivity rendering it an ideal wide band gap semiconductor for quantum electronics and optical devices useful under ambient and extreme conditions of high temperatures and intense radiation. Specifically, diamond has nitrogen-vacancy (N-V) defect centers with unusual characteristics making it attractive for these unique applications. One can control by microwave, optical signal, electric and magnetic fields the qubit states in N-V centers for quantum network, quantum memory and quantum sensing. Some of these applications require small nanometer or micrometer size diamond crystals containing preferably only one type of N-V defects for greatest sensitivity, individual addressability, and applicability. We are addressing these challenges by developing processing approaches to synthesize diamond single crystal arrays containing only one type of N-V defect centers by Microwave Plasma Enhanced Chemical Vapor Deposition (MPECVD). These results along with a brief overview on the promise of diamond for quantum applications will be presented and discussed.

InN and GaN Thin Films via Sputter Deposition

In this presentation, we demonstrate that reactive High-Power Impulse Magnetron Sputtering (HiPIMS) is an effective alternative for depositing high quality, epitaxial III-N thin films. In contrast to conventional direct current (DC) sputtering, HiPIMS provides the needed kinetic energy and ionization fraction to establish a sufficiently reactive environment to promote full nitridation. Specifically, the low duty cycle regime of HiPIMS provides access to kW/cm2 peak power densities without target degradation and thus produces dramatically increased metal reactivity. This unique capability set enables us to prepare high crystal quality epitaxial InN thin films with low out-of-plane mosaicity and electron mobilities in excess of 400 cm^2/Vs with a step-and-terrace microstructure when deposited on AlN-nucleated sapphire substrates. Equilibrium supersaturation models are useful for predicting the surface microstructure of these films by varying the V/III ratio or deposition temperature. In addition, homoepitaxial GaN thin films can be deposited at temperatures below 700°C with smooth surface morphologies characterized by c/2 steps and terraces. The presentation will discuss the relationships between sputtering parameters and III-N thin film crystal quality, surface morphology, and growth rate. Preliminary transport property measurements will be discussed.

SiC wafer cost reduction with remote epitaxy technology

Expanding SiC power semiconductor market requires lowering SiC wafer prices to reduce the majority cost of the device. In this talk, the author introduces the application of remote epitaxy technology through graphene to SiC alternative substrates with epi membranes. The difference between SiC and other remote epitaxy is that the growth temperature is as high as around 1873 in hot-wall chemical vapor deposition, which is a harsh environment for graphene survival. By replacing the carrier gas from hydrogen to argon, graphene etching was suppressed to peel off the membrane. The author believes this approach represents a significant step toward the fabrication of high quality and low-cost substrate with epi.

MOCVD growth of Gallium Oxide and Related Materials

β-Ga2O3 has recently attracted extensive interest in power electronics owing to its fundamental properties. Among the available epitaxial methods used to grow β-Ga2O3 films, MOCVD has proven suitable for producing high-quality epitaxial β-Ga2O3 films. The highest purity β-Ga2O3 films have been reported from MOCVD with record LT electron mobility exceeding 23,000 cm2/Vs and low~1013 cm-3 compensating acceptors [1]. This presentation will discuss recent progress in the MOCVD growth of high-quality β-Ga2O3 films and related materials. The use of various Ga precursors in realizing high-purity epitaxial films will be discussed. Process windows needed to realize high quality films using pure O2, H2O vapor, and N2O as oxygen sources will be presented. We will also examine the controllable doping of Ga2O3 using Si, Ge, N, and C dopants. This paper will also discuss the growth of high quality strained β-(AlGa)2O3/Ga2O3 and β-(InGa)2O3/Ga2O3 heterostructures and superlattices. Heavily doped (>1020 1/cm3) β-Ga2O3 and strained β-(AlxGa1-x)2O3/β-Ga2O3 heterostructures grown by MOCVD, and demonstration of record low resistance Ohmic contacts (ρc~10-7 Ωcm2) will be discussed. Finally, MOCVD's epitaxial maturity to grow high quality β-Ga2O3 films and its scalability for commercial scale production of these materials will be discussed. [1] G. Seryogin et al., APL 117 (2020) 262101.

Thin film Growth and Quantum Confinement of the Intrinsic Topological Superlattice Bi2-Bi2Se3

The spin-momentum locked surface state on Bi-Se topological insulators (TIs) have garnered a great deal of research as attractive prototypical devices for spin-orbit torque devices, magnetic field sensors, and polarized light detectors. To address the breadth of device needs, we demonstrate morphological, crystalline, and dimensionality control over TI Bi2-Bi2Se3 by varying the DC sputtering growth conditions. We uncover several interesting growth trends, 1) nano-platelet nucleation with an island growth mode until a pressure-dependent critical thickness and transition to layer-by-layer growth, 2) an amorphous to crystalline transition as a function of temperature with competing lowest energy growth orientations, 3) an atomically smooth, to highly faceted morphology transition as a function of pressure, and finally, 4) quantum confinement in nano-platelet with a dramatic increase in optical bandgap and softening of Raman modes. We observe high-quality van der Waals layers with twinning via transmission electron microscopy. With local scanning tunneling spectroscopy, we reveal a linear LDOS around the Fermi-level in excellent agreement with angle-resolved photoemission spectroscopy. This work capitalizes on the commercial scalability of sputtering to develop key growth trends that enable advancements in a variety of technologies exploiting topological surface states.

Charge and strain tunable magnetism at the MnTe/InP interface

Engineering interfaces in materials systems with reorientable spins networks is of significant technological interest for applications in spintronics. At the same time, the promise of antiferromagnetic spintronic has resulted in a renewed intensive study of antiferromagnetic materials with large spin-orbit coupling. MnTe is a layered hexagonal, room-temperature antiferromagnetic (TN ≈ 310 K) semiconductor (energy gap of order 1 eV) with a NiAs structure of alternating layers of Mn and Te. The anisotropy axis and magnetic ordering of MnTe is sensitive to charge doping and as such opens the door to tailoring interlayer magnetic interactions in this layered materials system. Here, we present results on epitaxially grown MnTe/InP(111) heterostructures. Using detailed magneto-transport studies combined with magnetometry data, we observe emergent ferromagnetic behavior at the interface driven by charge transfer and strain. This interfacial ferromagnetic layer is strongly coupled to the switching behavior of the antiferromagnetic layer. These results provide us a potential mechanism of tuning antiferromagnetic ordering for applications in high-speed, next-generation spintronics.

Probing the electron-phonon dynamics around the band gap of amorphous silicon telluride

Amorphous chalcogenide alloys are key materials for data storage and energy scavenging applications due to their large non-linearities in optical and electrical properties. Silicon telluride (SiTe) is an example of these alloys because it undergoes a solid-state phase transformation upon application of a sufficiently large electric field, commonly known as threshold switching. Here, we investigate the impact of this phase transformation on electrical properties by monitoring carrier lifetimes at and around the transformation. We implement an ultrafast near-infrared pump- probe spectroscopy technique that allows for direct monitoring of electronic and vibrational energy carrier lifetimes in SiTe. We probe the reflectivity response of this alloy after pump heating in the energy range of 0.95 to 2.07 eV. SiTe demonstrates the direct valence mobility edge to conduction mobility edge transition at 1.24 eV. Besides, we identify a hole trap state at 1.08eV, which is near the valence band edge, indicating that the flow of electrons in SiTe is possible under the threshold voltage. The electrons of the low energies relaxes faster than the free electrons of bandgap energy, confirming the presence of trap states in SiTe and its conductive nature under the threshold voltage. These findings open the way of tuning the property of this alloy, which will lead to building new electronic devices with higher efficiency.

3D structure–property correlations of emerging ferroelectric materials revealed by tomographic atomic force microscopy

With ever-increasing complexities in the structure and design of advanced functional materials--encompassing single crystals and epitaxial films, heterostructures and superlattices, polycrystallinity, nanocomposites and mesoporosity, and 2D materials and 3D patterning--there is a strong demand for imaging and analysis methods with volumetric property mapping capabilities at the nanoscale. After a brief review of the tomographic atomic force microscopy approach (T-AFM), results from several emerging ferroelectric materials are presented. These include the 3D complex domain structures in BiFeO3-CoFe2O4 nanocomposites, polycrystalline HfO2 thin films, and oxide-free 2D van der Waals ferroelectrics. Thickness scaling of their functional properties are also uniquely accessible, including the polarization magnitude and preferential orientations as well as ferroelectric switching. The novel perspective provided by this multi-dimensional imaging approach, providing insight into correlations between local properties and microstructure, strain, interfaces, and thickness dependences, thereby advances our fundamental understanding and informs practical designs for future ferroelectric materials and devices.

Homogeneously Miscible Fullerene inducing Vertical Gradient in Perovskite Thin-Film toward Highly Efficient Solar Cells

Fullerene-based n-type charge-collecting materials have emerged as a solution for high-performance perovskite solar cells. However, their application to perovskite solar cells is limited in the device architecture and only a small amount of fullerene additives have been introduced to the device system, because of the immiscibility of the fullerene species with polar solvents. To overcome this, triethylene glycol monomethyl ether chain-attached fullerene derivatives are synthesized and applied to normal-type perovskite solar cells. The newly synthesized fullerenes exhibit excellent solubility in polar solvents. A novel approach to introducing miscible fullerenes into perovskite devices and inducing a favorable vertical gradient is proposed. Forming an overcoat on an electron-transporting layer and waiting for a few minutes, the fullerene derivatives progressively permeate into the fullerene-doped perovskite active film. By fabricating perovskite solar cells combining direct mixing, overcoating, and waiting techniques, a remarkably high device efficiency of 23.34% is achieved. The high performance is attributed to the fullerene additives with a vertical gradient passivating the perovskite defect sites effectively and the overcoat enhancing the charge transfer.

Band-gap engineering of ABO3 (A = Ba) perovskites by isovalent B-site substitution

Solar cells made of inorganic perovskite absorbers achieved recently up to 8 % power conversion efficiency. Still, current studies deal with very few suitable perovskite oxides for photovoltaic applications even though the field offers a wide range of material combinations. However, within oxides it is challenging to find narrow band-gap materials. In this contribution, we show how the band-gap can be modified systematically by introducing Cu2+ as well as Ta/Nb5+ or W/Mo6+ on the B-site of the perovskite. The impact of these substitutions on important semiconducting properties, e.g. optical and electrical characteristics, is shown. ABO3 perovskites containing multiple cations on the B-site are synthesized by solid state reactions and several hundred nanometres thick thin films are grown onto transparent conductive oxides by pulsed laser deposition. Fundamental properties are characterized by Rietveld refinement of diffraction data. Microstructure and composition are analyzed by SEM and EDX, respectively.

Entropy as a control over defects, doping, and phase in strongly correlated materials

Disorder is an important aspect of correlated oxide systems - it can be used to manipulate superconductivity, magnetic ordering, and design degeneracies. Phase order classically relies on uniformity, with disorder and defects resulting in lower ordering temperatures and prevention of long-range percolation. However, in a chaotic limit, disorder can be treated as a parameter which drives the local microstates into global order. We demonstrate that this seemingly counterintuitive statement is not only true but that it simplifies the prediction of predominant phase in high entropy oxides. As a result, we design, grow and characterize high entropy perovskite oxide films demonstrating the utility of this predictive materials approach. We present these theoretical and experimental results on two classes of single crystal epitaxial films. First, in exploring magnetism of the high entropy ABO_3 perovskite $La(Cr_{0.2}Mn_{0.2}Fe_{0.2}Co_{0.2}Ni_{0.2})O_3$. Second, in a case of extreme A-site cation disorder in $(Y_{0.2}La_{0.2}Nd_{0.2}Sm_{0.2}Gd_{0.2})NiO_3$, whose parent ternary oxides each have a large range of electronic and structural phase transition temperatures. These results suggest cation variance, such as that accessible only in high entropy oxides, can be a critical order parameter in the design of correlated oxides, providing continuous tunability to charge, electronic, structural, and magnetic orderings.

Atomically Engineered Thin Films and Devices

Thin film materials and heterostructures play a key role in modern technology including electronics and photonics. Atomically precise engineering of thin film materials enables unprecedented control of their structure and properties, bringing exciting opportunities to future materials and devices. Transition metal dichalcogenides (TMDs), which form three-atom-thick monolayers with van der Waals surfaces, provide an ideal material platform with diverse electrical and optical properties for thin film engineering in the atomically thin limit. In this talk, I will discuss three key synthesis challenges for realizing atomically engineered thin films and superlattices with atomically thin TMDs. Firstly, I will discuss how high-performance monolayer TMD films can be synthesized with wafer-scale uniformity. Furthermore, I will discuss how dissimilar TMDs (e.g. tungsten disulfide and tungsten diselenide) can be integrated laterally in the monolayer plane to form superlattices without dislocations, despite a large 4% lattice mismatch. Lastly, I will discuss how various TMD monolayers can be stacked vertically to form superlattices with designed properties. These scalable synthesis capabilities will further enable novel atomically engineered materials that hold great potential for future ultrathin electronics.

Designing High-Tc Superconductors with BCS-inspired Screening, Density Functional Theory and Deep-learning

Recent advances in first principles calculations and machine learning techniques allow for a systematic search for phonon-mediated superconductors. We develop a multi-step workflow for the discovery of conventional superconductors, starting with a Bardeen-Cooper-Schrieffer inspired pre-screening of 1736 materials with high Debye temperature and electronic density of states at the Fermi-level. Next, we perform electron-phonon coupling calculations for 988 of them to establish a large and systematic database of BCS superconducting properties. Using the McMillan-Allen-Dynes formula, we identify 112 dynamically stable materials with transition temperatures, $T_c > 5$ K. In addition, we analyze trends in our dataset and individual materials including MoN, VC, VTe, KB_6, Ru_3NbC, V_3Pt, ScN, LaN_2, RuO_2, and TaC. Finally, we demonstrate that deep-learning models can predict superconductor properties, including the Eliashberg function, thousands of times faster than direct first principles computations. We apply the trained model on the crystallographic open database and pre-screen 8293 candidates for further DFT calculations.

Search for Advanced Electric Conductors – an Emerging Challenge

The generation, distribution and control of electric power is fundamental to every electronic industry, and power levels range from even milliwatts up to 10's of Gigawatts. Essential elements of electric power systems include electric conductors, and for aerospace they are the major weight percentage (surprisingly, by-far) of aircraft electric components and systems. For microelectronic integrated circuits, the heat loss of Cu/Al interconnects has stopped Moore's law of progress, and reduce the system-level switching speed by > 100x. And for Quantum Information Technology (QIT), as systems scale to larger size, the quantum interconnect (QuIC) bottleneck is imminent, and emerging as a grand challenge. The solutions for these technology challenges is similar, to find/develop electric conductors that have superior electrical conductivity to standard materials, including Cu and Al. And for aerospace applications, there is an extra need to greatly reduce the conductor weight. This talk will introduce and review these topics, and potential materials classes that can be studied to increase the mass-specific electrical conductivity, for wide ranges of temperature needed from 50 mK to > 1000 K. Potential classes of materials to consider include carbon allotropes, delafossites, Weyl metals, and other.
Acknowledgments: support by AFOSR LRIR #18RQCOR100, and the Aerospace Systems Directorate.

Monolayer superconductivity and tunable topological electronic structure at the Fe(Te,Se)/Bi2Te3 interface

The interface between two-dimensional topological Dirac states and an s-wave superconductor is expected to support Majorana bound states that can be used for quantum computing applications. Realizing these novel states of matter and their applications requires control over superconductivity and spin-orbit coupling to achieve spin-momentum locked topological surface states which are simultaneously superconducting. While signatures of Majorana bound states have been observed in the magnetic vortex cores of bulk FeTe0.55Se0.45, inhomogeneity and disorder from doping makes these signatures unclear and inconsistent between vortices. Here we will present superconductivity in monolayer FeTe1-xSex grown on Bi2Te3 by molecular beam epitaxy and investigate the interfacial spin and electronic structure with photoemission spectroscopy. We will describe how to systematically tune the interfacial topological and superconducting electronic structure through control over film growth conditions. We will show how reducing Se doping not only reduces disorder but improves the spin-momentum locking of the interfacial topological states which is important for Majorana interrogation and potential applications.

The pairing symmetry in quasi-one-dimensional superconductor $Rb_2Mo_3As_3$

The $A_2Mo_3As_3$ (A=K, Rb, Cs) compounds are made of assembled Mo_3As_3 chains and are thus potential realizations of quasi-one-dimensional metals. Various experiments on these materials hint to the Tomonaga-Luttinger liquid (TLL) physics from which superconductivity with a relatively high critical temperature of T_c = 10.4 K emerges. However, TLL and its relevance to superconductivity are not yet resolved for $A_2Mo_3As_3$ due to their multiband nature. Here we report a combined ^{75}As nuclear quadrupole resonance (NQR), ^{87}Rb nuclear magnetic resonance (NMR) and muon spin relaxation (μSR) study of $Rb_2Mo_3As_3$ powder samples. The ^{87}Rb spin-lattice relaxation rates show a characteristic power-law temperature dependence over a broad temperature range, which is a hallmark of the TLL dominated by the two quasi-one dimensional bands. Detailed analysis revelas that $Rb_2Mo_3As_3$ is in the unusual TLL regime of attractive interactions, which opens intriguing possibilities for the unconventional pairing symmetries of the emerging superconducting state. To address the superconducitng pairing symmetry below T_c we employed transverse-field μSR. The temperature dependence of the muon relaxation rate, σ, is in the superconducting state due to the field distribution created by the vortex lattice and was compared against s-, p- and d-wave scenarios. The penetration depth λ=669 nm and the coherence length ξ = 3.4 nm were also determined.

Competing Incommensurate Spin Fluctuations and Magnetic Excitations in Infinite-Layer Nickelate Superconductors

The recently discovered infinite-layer nickelates show great promise in helping to disentangle the various cooperative mechanisms responsible for high-temperature superconductivity. However, lack of antiferromagnetic order in the pristine nickelates presents a challenge for connecting the physics of the cuprates and nickelates. Here, by using a quantum many-body Green's function-based approach to treat the electronic and magnetic structures, we unveil the presence of many two- and three-dimensional magnetic stripe instabilities that are shown to persist across the phase diagram of $LaNiO_2$. Our analysis indicates that the magnetic properties of the infinite-layer nickelates are closer to those of the doped cuprates which host fluctuating, intertwined ground states rather than the ordered undoped cuprates. The computed magnon spectrum in $LaNiO_2$ is found to contain an admixture of contributions from localized and itinerant carriers. The theoretically obtained magnon dispersion is in accord with the results of the corresponding RIXS experiments. Our study gives insight into the origin of inhomogeneity in the infinite-layer nickelates and their relationship with the cuprates.

Non- and near-contact Andreev reflection for quantitative microscopy of superconductors with atomic resolution

Andreev reflection is an established method to probe the existence of superconductivity, and, crucially, the symmetry of the order parameter. However, the popular point contact method for the measurement of Andreev reflection (PCAR) necessitates formation of good mechanical contact, which severely limits spatial resolution and introduces complicated interpretation, particularly in the important case of unconventional and multi-band superconductors. In this talk, we will present our recently developed non-contact Andreev reflection (NCAR) technique - a new experimental approach to quantify Andreev reflection without mechanical contacts. The technique utilizes the fundamental connection between the physics of the scattering process and the strength of exponential non-linearity of the tunneling current, and therefore adopts the renormalized tunneling current decay rate as the experimental observable. Due to this choice of the observable, NCAR enables robust theory-experiment comparison, detailed quantification of the presence and symmetry of the order parameter, and even quantification of the broadening mechanisms that compete with superconductivity.

Toward control and manipulation of superconducting vortex lattices from nano to mesoscales

In this talk, we will discuss two experiments that demonstrate direct control over vortex positions in FeSe superconductor. First, we observed that twin boundaries in the FeSe trap a high density of vortices and align vortex configuration in FeSe. The alignment effect causes various vortex configurations such as one-dimensional and square lattice – both with ordering qualitatively different from the commonly observed vortex glass. The analysis of these interactions also reveals further clues regarding vortex dynamics, inter vortex interaction, and vortex pinning. Second, we found that the vortex shape can be controllably varied with the imaging conditions within STM – the effect which we attribute to direct evidence of vortex manipulation. Finally, we have explored extreme conditions of STM imaging where vortex positions can be locally manipulated. Our results give hints to the understanding of vortex manipulation in STM geometry toward quantitative nanoscale probe of vortex dynamics and a platform to explore vortex manipulation in the context of topological quantum computing. Work supported by the U.S. Department of Energy, Office of Science, Materials Sciences and Engineering Division. Experiments were carried out as part of the user project at the Center for Nanophase Materials Sciences, Oak Ridge National Laboratory, which is a US Department of Energy Office of Science User Facility.

Tunable Electronic States and Instabilities in PbSnTe Heterostructures

Combination of broken inversion symmetry and spin-orbit coupling gives rise to a wide range of exotic superconducting states, such as mixed-parity superconductivity, superconducting Weyl state, and superconducting diode effect. Incipient ferroelectrics e.g., PbTe, are near a polar instability. Pb1-xSnxTe demosntrates a highly tunable superconductivity and ferroelectricity which are dependent on carrier density and Sn(Pb) concentration. Furthermore, a topological phase transition is expected at x~0.4, which inverts the conduction and valence bands for Sn rich compounds. I will report on our recent growth efforts of high quality Pb1-xSnxTe heterostructures, using a chalcogenide molecular beam epitaxy (Veeco 930) near the topological phase transition (x~0.4). The cross-section high-angle annular dark-field (HAADF) imaging in scanning transmission electron microscopy (STEM) shows abrupt interfaces. X-ray diffraction demonstrates single oriented films. The sheet resistance v.s. temperature demosntrates a metallic-like behavior, dR/dT >0, in doped samples extending to 2 K. The Hall carrier density was measured at various temperatures, resolving the carrier mobility. The carrier mobility is inversely proportional to doping concentration. We report on superconductivity in electron and hole doped Pb1-xSnxTe heterostructures and its relation to neighboring ferroelectric and topological phase transitions.

Magnetic imaging of domain walls and surface magnetism in antiferromagnetic topological insulators

Layered antiferromagnets with topological band structure are promising platforms to host many interesting topological phenomena. Thus, it is imperative to visualize domains or domain walls in these topological antiferromagnets. Despite many decades' efforts, it remains a grand challenge to visualize antiferromagnetic domains or domain walls in magnetic field. Here I will present magnetic imaging of domain walls in antiferromagnetic topological insulator MnBi2Te4 family using cryogenic magnetic force microscopy (MFM). Our MFM studies reveal enhanced magnetic susceptibility inside the domain walls due to the winding of the antiferromagnetic order parameter. In addition, our MFM data indicates the A-type antiferromagnetic order persists to the surface of MnBi2Te4 and natural superlattice compound MnBi4Te7, which shed new light on the mystery of gapless topological surface states observed by high resolution angle-resolved photoemission spectroscopy. The robust A-type order is further corroborated by the surprising discovery of the long-sought surface spin-flopand surface spin-flip transitions, respectively. Our results pave the way for exploration of domain walls or surface magnetism in topological antiferromagnets using magnetic imaging.

The electronic detection of quantum spin fluctuations in geometrically frustrated magnet heterostructure

While the rise of quantum computers may one day help solve complex problems and deliver information with unhackable security, there is lack of a material platforms for scalable realization of quantum technologies. For instance, the most interesting magnetic property of the celebrated quantum spin liquids (QSLs) is the possibility of quantum mechanical encryption and transport of information, protected against environmental influences. Despite extensive studies on QSLs, they are still far away from applications. One obstacle is that most of the studied QSLs are insulators and electronically inert, which is incompatible with an electrical circuit that relies on moving charge carriers. The grand challenge is to find a way to convert the entanglement information into mobile charge signal by "metallizing" quantum magnets. In this talk, I will introduce a unique approach by designing and synthesizing new heterostructures based on geometrically frustrated magnets to address this obstacle. I will give two examples: (i) $Dy_2Ti_2O_7/Bi_2Ir_2O_7$ heterostructure in which the transport data measured on the $Bi_2Ir_2O_7$ film reflects the Kagome spin ice broken phase transition of the underlying $Dy_2Ti_2O_7$ crystal; (ii) $Yb_2Ti_2O_7/Bi_2Ir_2O_7$ heterostructure in which the abnormal magnetoresistance of the $Bi_2Ir_2O_7$ film strongly correlates with the strong quantum spin fluctuations in $Yb_2Ti_2O_7$.

Synthesis and Design of Low-Dimensional Chalcogenides for Infrared Nonlinear Optics

Acentric chalcogenides are prime candidates for nonlinear optically-active (NLO) materials in the infrared (IR) due to their relatively weak interatomic forces, leading to wide optical transparency regions and second-harmonic generation (SHG) in the near to mid-IR. Alkali chalcoarsenates, $AAsQ_2$ (A = Li, Na, K; Q = S, Se), are high-performing materials with SHG responses orders of magnitude higher than industry standard materials, $AgGaS_2$ and $AgGaSe_2$. These have primarily been studied by changing the cation to observe changes in dimensionality, and subsequently the NLO behavior. Here, we report anionic tuning in the isostructural compounds $LiAsS_{2-x}Se_x$ (x = 0, 0.25, 1, 1.75, 2) which crystallize in the noncentrosymmetric space group Cc. These materials are one-dimensional (1-D) $[AsQ_2]^-$ (Q = S, Se, S/Se) chains consisting of corner-sharing AsQ_3 trigonal pyramids with charge-balancing Li^+ atoms interspersed between the chains. As the composition becomes Se-rich, the materials become phase-matchable, SHG response increases, and the laser-induced damage threshold (LIDT) decreases. We grew single crystals of particularly promising members and found one of the highest nonlinear coefficients reported thus far (d_{eff} < 400 pm/V). Anionic tuning is an underutilized strategy to methodically modulate the chemical and optical properties of materials of interest and is used here with great success.

The field induced quantum spin liquid behavior in a spin-1/2 honeycomb magnet $Na_2Co_2TeO_6$

The Kitaev model is an exactly solvable quantum spin liquid model of two-dimensional magnets, and the itinerant Majorana excitations was observed in the low-energy behaviours, hence, it is one of the frontiers in quantum condensed matter physics. Recently, 3d-oxide, $Na_2Co_2TeO_6$, was reported as a 3d-based Kitaev candidate. With the measurements of the in-plane magnetic torque, Muon spin resonance, magnetization, and heat capacity, a magnetic field vs. temperature phase diagram was built up, and a series of states have been observed, including the zigzag ordered phase, an intermediate disordered phase, and the polarized trivial phase. The angular dependences of the magnetic torque and susceptibility indicate that the crystalline point group symmetry is restored due to the quantum fluctuations. In addition, the inelastic neutron scattering measurement has been applied to study the spin dynamics on the zigzag ordered state and the disordered state, and the spectra well agree with the simulation of the effective K-J_1-Γ-Γ′-J_3 model.

Three-stage magnetic phase transition in ultrahigh-quality 2D CrSBr magnet

2D magnets down to the atomic limit have received widespread attention nowadays due to their significance in both fundamental research and potential applications in spintronic devices. New materials with preferably in-plane magnetic easy axis, good crystal and exfoliation qualities, high air stability, and interfacial properties will be rather attractive and rewarding in the 2D magnets studies. CrSBr emerges as such a candidate where single layer of CrSBr shows a ferromagnetic order and these FM layers couple antiferromagnetically along the c-axis. In this talk, we will present ultrahigh-quality air-stable CrSBr crystals synthesized by us. The high single crystallinity and spatial homogeneity have been thoroughly evidenced at length scales from sub-mm to atomic resolution by X-ray diffraction, second harmonic generation, and scanning transmission electron microscopy. Our ultrahigh-quality CrSBr can further be exfoliated down to monolayers and bilayers easily. More importantly, we show thermodynamic evidence of a stage-by-stage development of three magnetic orders upon cooling, at 185K, 156K, and 132K respectively, which is also corroborated with the magnetization and transport results. The details of the results and their implications will be discussed.

Transport spin polarization of Pt/ferromagnetic insulator bilayers

It has been shown that the anomalous Hall effect can be observed in platinum films deposited on a ferromagnetic insulator which implies the existence of a magnetic proximity effect [1]. One of the key features of ferromagnetic metals is a net spin polarization of the carriers at the Fermi energy which enable the functionality of spintronic devices. In this presentation we show that spin polarized carriers are indeed present in Pt films of various thicknesses that were deposited on magnesium aluminum ferrite (MAFO) using point contact Andreev scattering [2]. The results were obtained by extracting conductance vs. voltage data from I-V curves taken through contacts formed by driving a sharpened superconducting Nb tip into the Pt/MAFO samples. The resulting spectra were then analysed using a modified BTK theory of supercurrent conversion at a normal/superconductor interface to obtain the values of the transport spin polarization. References Z. Wang, C. Tang, R. Sachs, Y. Barlas, and J. Shi, Phys. Rev. Lett. 114, 016603 (2015). R. J. Soulen, Jr., J. M. Byers. M. S. Osofsky, B. Nadgorny, T. Ambrose, S. F. Cheng, P. R. Broussard, C. T. Tanaka, J. Nowak, J. S. Moodera, A. Barry, and J. M. D. Coey, Science 282, 85 (1998).

Imaging and Manipulation of Magneto-Acoustic Interactions in Magnetostrictive Devices

Magnetic materials play a crucial role in memory, computing, and microwave processing devices due to reconfigurable domains, GHz resonance, and coupling to controllable variables such as magnetoelectric fields, interfaces, and strain. However, conventional devices often rely on current induced magnetic fields or high current density architectures which lead to size, power, or heating issues. Recently an alternate coupling mechanism, the magneto-acoustic interaction, has gained interest. Surface acoustic wave (SAW) devices driven by voltage controlled interdigitated transducers provide a method of driving low power acoustic interactions in magnetic films. In magnetostrictive materials, these interactions can drive magnetic excitations including resonance and domain motion, and devices have exhibited novel behavior including nonreciprocity of SAW phase and amplitude. In this talk, I will discuss our work on the development of these magneto-acoustic devices. This will focus on a discussion of the role imaging techniques, including MOKE, Brillouin light scattering, and laser interferometry, play in our characterization of the devices and understanding of the acoustic and magnetic behavior. I will also touch on the role that material choice, growth condition, and acoustic wave shaping play in device performance and comment on materials development and fabrication challenges for this field.

Probing Ion Migration in ABX3 Perovskite Compounds: Five Fallacies of Simulations

The migration of the constituent ions of the ABX3 perovskite structure is attracting ever increasing attention because of its relevance for a variety of applications. In some cases, the rate of ion migration should be increased, for example, in materials based on LaGaO3 or (Na,Bi)TiO3 for fuel cell electrolytes, while in other cases, BaTiO3 for capacitors or CH3NH3PbI3 (MAPbI3) for photovoltaic devices, it should be minimised. Simulation studies have a central role to play in interpreting and explaining experimental transport data and providing insights at the atomic scale. A variety of simulation techniques have been employed to examine ion migration in perovskites to date. These techniques involve molecular static or molecular dynamics calculations, and employ classical pair potentials, reactive force-fields or quantum-mechanical calculations. Regardless of the technique, the results of simulations are often regarded as "scientific truth". In this contribution, taking ion migration in BaTiO3, CaTiO3, LaMnO3 and MAPbI3 as examples, we draw attention to problems that may arise when using static calculations to obtain activation barriers and to issues with molecular dynamics simulations of diffusion. We demonstrate that a far more critical approach to simulation results in the literature is required.

Tuning functional properties of complex oxides via defect engineering

Defects and their atomic arrangements play a significant role in physical properties of complex oxides. Therefore, fundamental understanding of how to tune and control these defects is essential for advancing the functionalities of complex oxides. In this talk, I will present our recent effort in tuning and controlling the defect formation and migration during materials synthesis and processing processes. The first part of my talk focuses on oxygen-deficient brownmillerite (BM)-structured $SrFeO_{2.5}$ (BM-SFO) with ordered oxygen vacancy channels (OVCs). By tuning the thin film deposition parameters or substrates, I show we can control the $SrFeO_{2.5}$ thin films with different orientations of OVCs. By using in situ environmental TEM, we observed directional oxygen transport from the reducible substrate side to $SrFeO_{2.5}$ thin films during TEM imaging. In addition, we found that oxygen ions can be guided to transport along two distinct pathways in differently oriented $SrFeO_{2.5}$ thin films, resulting in structurally and electronically different polymorphs of $SrFeO_{2.75}$ as intermediate states. In the second part, I will present the epitaxial growth studies of $LiCoO_2$ with or without overlayers to understand how the defects form and then affect the Li ions transport processes. Both parts highlight the critical roles of defects, which are of vital importance in the design of functional materials and energy storage devices.

Effects of Compositional Phase Separation on Mass Transport in Resistive Memory

Resistive random access memory (ReRAM) is an important technology for storage memory, in-memory computing, and neuromorphic computing. ReRAM switches and stores analog resistance states through the migration of oxygen vacancy point defects in transition metal oxides, which form or breaks conducting filaments. Understanding the mechanism of oxygen vacancy migration is crucial towards understnading switching mechanisms. In resistive memory, mass transport is universally modeled using Fick's First Law, which states that oxygen migraties from high to low concentrations, under the assumption that the metal and oxygen form an ideal solid solution. In this work, we utilize a combination of materials characterization, device measurements, and physical modeling to show that this ideal solution assumption is inaccurate for the well-studied ReRAM material tantalum oxide (amorphous). Instead, the tantalum oxide undergoes compositional phase separation due to nonideal defect interactions. This phase separation causes the filaments to be thermodynamically stable because the oxygen vacancies are attracted to other oxygen vacancies. Our work shows the importance of properly accounting for defect-defect thermodynamic interactions in describing memristive devices.

Anion intermixing and transport during oxide thin film deposition revealed with isotopic tracers

Thin film deposition from the vapor phase is a key technology that enables both commercial fabrication for a wide variety of applications and the research and development of new materials and devices for next-generation technologies. Despite its importance, a comprehensive, predictive understanding of the thin film deposition process is still lacking. Thin film morphology during well-controlled, non-energetic deposition by molecular beam epitaxy (MBE) is understood as a competition between surface and interface energies, tempered by adatom mobility, leading to the well-known layer-by-layer (Frank-van der Merwe), layer-island (Stranski-Krastanov), and 3D island (Volmer-Weber) growth modes. However, these models typically ignore intermixing and diffusion. Here we present quantitative experimental evidence of anion intermixing over long length scales during the deposition of epitaxial Fe_2O_3 and Cr_2O_3 films that incorporate ^{18}O tracers. Intermixing events suggested by molecular dynamics simulations are examined to reveal that adatoms on the film surface act to “pull up” subsurface O and cations. Subsequent ring-like rotation mechanisms involving both adatom and subsurface anions then facilitate their mixing. These intermixing mechanisms may also be operant during other surface-mediated processes such as corrosion.

Tailoring charge transport in mixed conducting oxide thin films via crystallization

The ability of mixed ionic/electronic conducting (MIEC) oxides to exchange oxygen with the environment and transport ionic and electronic species motivates their use in solid-state electrochemistry. Chemical strains from changes in oxygen stoichiometry can be exploited to actuate hierarchical structures through low-temperature crystallization of thin films, transforming key properties: Our previous work has reported this low-temperature processing route leads to enhanced surface exchange kinetics and in-plane electronic conductivity in crystallized titanate perovskite films. Our current effort explores the interplay between crystallinity and charge transport in amorphous-grown $La_{0.7}Sr_{0.3}Ga_{0.6}Fe_{0.4}O_{3-x}$ thin films. This work aims to decouple the ionic contribution from the total conductivity through electrical measurements with blocking electrodes. In-situ, simultaneous grazing incidence X-ray diffraction and ac-impedance spectroscopy measurements corroborated by ex-situ dc polarization measurements reveal the ratio of ionic conductivity to the total conductivity (transference number) can be tailored through the degree of crystallinity. As deposited, amorphous LSGF films exhibit mostly ionic conduction. With increasing crystallinity, the films transition into mostly electronic conductors. Implications for low-to-intermediate temperature solid-state electrochemical applications will be addressed.

Modification of Schottky barrier and electrical properties of oxide ceramics by dislocations

For decades, controlling point defects and interfaces have been the main method of engineering the electrical and photoelectrical properties of inorganic semiconductors. However, due to the brittle nature of ceramics, dislocations that are typically introduced in the material by mechanical deformation have been mostly disregarded. Recent advances in tunability of functional properties of ceramics, such as ferroelectric, thermal, and electrical properties, have attracted extensive attention. The present work investigates the effect of dislocations on the electrical properties of p-type and n-type SrTiO3. We show that large volumes with dislocation densities higher than 2×1013 can be produced by cyclic loading and subsequent sample thining without the formation of any cracks. With the help of electrochemical impedance spectroscopy and DC electrical measurements, we illustrate that the Schottky barrier in the n-type SrTiO3 can be reduced by a factor of seven, while the bulk resistivity of the p-type one can be increased by a factor of three.

Dislocation-tuned electrical conductivity in functional ceramics

The functional properties of ceramics, e.g., electrical conductivity, are usually tailored by designing point defects and interfaces. Dislocations as heavily charged nanoscale one-dimensional line defects have so far been underappreciated as a means to tune the electrical conductivity but are finding increasing attention today. Here, based on electronic and ionic conducting functional ceramics, we illustrate that mechanically introduced dislocations possess the ability to modify the electrical conductivity significantly. This will be rationalized based on results from investigations on rutile (TiO_2) and yttria-stabilized zirconia (YSZ). The impact of dislocations on these functionally important ceramics depends on a convolution of dislocation character, core properties, possibly existing space charges, and mesoscopic structure. The control of dislocation generation is nevertheless challenging as ceramics are usually not plastically deformable. We combine the understanding of mesoscopic dislocation structure and its behavior at elevated temperatures. The interaction of dislocations and bulk properties is thus both complex and exciting as it allows for conductivity modification possibilities in functional ceramics.

Atomistic Mechanism of Lithium Ion Diffusion at Grain Boundaries and Surfaces in Li-Metal Oxides

The properties of lithium ion battery cathodes and electrolytes strongly depend on the diffusion of lithium ions from the surfaces and across grain boundaries (GBs) during charge/discharge process. Then, direct visualization of lithium site is required to understand the mechanism of the diffusion behavior of lithium ions. In this study, aberration (Cs) corrected STEM were applied to directly observe the surface atomic structures of olivine Li_xFePO_4 and the grain boundary (GB) atomic structures of lithium lanthanum titanate (LLTO). It was found that thin intermediate $Li_{2/3}FePO_4$ phase was formed at the $FePO_4$/ Li_xFePO_4 interface, and lithium diffusion occurred by dragging the intermediate phase from inside to the surface. The mechanism of the lithiation/delithiation from and to the surface will be discussed based on the observation results. Concerning LLTO, different types of GBs were prepared by fabricating the bicrystals, and their structures were observed by Cs corrected STEM and atomic force microscopy (AFM). Then, the charge states, Li-ion conductivities, atomic and electronic structures were systematically and quantitatively investigated. It was found that the $\Sigma 5$ GB has no significant influence on Li-ion conductivity, but the $\Sigma 13$ GB shows the significant reduction of the Li-ion conductivity. The atomistic mechanism will be discussed in this presentation.

Defect-focused, high-throughput screening of ionic & mixed conductors

Accelerated discovery of solid-state ionics benefits from rapid evaluation of ionic conductivity and catalytic activity across wide composition spaces. Descriptor filtering of databases may identify promising compositions, but the lack of defect thermodynamic data hinders the ability to directly select for strongly defect-mediated properties like ionic conductivity. Also, accurate calculations of defect populations are too costly to implement in high-throughput screening. Instead, we developed a new descriptor filter invoking a "proximal stoichiometry" feature that can be calculated from Materials Project data to identify compositions with dominant Frenkel disorder, as candidate battery electrolytes. Subsequent in-depth analysis of air-stable candidates identified in Li-Bi/B-O systems shows near-pure Li-ion conductivity. Conversely, for fast experimental screening, methods for measurement of defect-mediated properties across composition libraries can be developed for combinatorial films. We have instituted a high-temperature 2D-optical transmission relaxation method to screen electrochemical reaction kinetics (e.g., oxygen surface exchange coefficient, k) simultaneously across each point of a mixed conducting film. The technique leverages the relationship between defect concentration and optical absorption. Initial application to study the impact of composition on k in titanate perovskites will be discussed.

Determination of point defects with Dark Field X-ray Microscopy (DF-XRM)

Oxygen vacancies are one of the most common such defects in metallic oxides, and have been reported to play a crucial role in the physical functionalities. In view of the enormous relevance of oxygen vacancies in ferroelectric materials, we study their nature, distribution and concentration by using Dark-Field X-ray microscopy (DF-XRM), which overcomes the surface-scale and indirect limitations usually found in advanced spectroscopy and microscopy methods and classical chemical characterization by probing directly the elastic deformation caused by buried point defects in oxide materials. We carry out the analysis of the experimental results through a comparison with simulated 3D multiscale models. This method consists of a combination of atomistic simulations from DFT and finite elemental analysis in COMSOL Multiphysics. This approach allows us to recreate 3D maps of the local strain produced by the oxygen vacancies down to the resolution of our instrument. We then use the obtained strain field to simulate Dark-Field X-ray microscopy images pixel by pixel. The support of simulations allows us to characterize point defect distributions in experimental data. This is the first time that DF-XRM is used to study point defect and more importantly that a multiscale method is developed in order to support the experimental data.

Oxygen vacancy migration and impact on high voltage DC polarization in BaTiO3-Bi(Zn,Ti)O3

BaTiO3-Bi(Zn,Ti)O3 (BT-BZT) is a relaxor dielectric with high energy density and high resistivity at elevated temperatures. While BT-BZT has shown attractive properties, there have not been any studies of its susceptibility to long-term degradation in resistance under high electric field (DC lifetime). Oxygen vacancies unintentionally added by impurities or formed from cation vacancy formation during high temperature sintering are often suggested to limit DC lifetime in dielectrics. Thermally stimulated depolarization current (TSDC) is used here to probe the oxygen vacancy migration and relative concentration. The oxygen vacancy concentration is modulated using thermal anneals and the impacts on TSDC response and DC lifetime are characterized. The role of microstructure including grain boundaries and space charge will be discussed. Sandia National Laboratories is a multimission laboratory managed and operated by National Technology & Engineering Solutions of Sandia, LLC, a wholly owned subsidiary of Honeywell International Inc., for the U.S. Department of Energy's National Nuclear Security Administration under contract DE-NA0003525.

Fundamental Study of Defect Migration under Burn-in Conditions for High Reliability in Multilayer Ceramic Capacitors

This study involves a detailed investigation of base metal multilayer ceramic capacitors. The failure mechanisms are considered to be predominantly related to oxygen vacancy migration under bias. The kinetics of this process can be monitored with highly accelerated lifetime test and other electrical characterization, such as thermally stimulated depolarization current and impedance spectroscopy. To better screen high reliability capacitor components, there are protocols where components are exposed to various field and temperature conditions relative to their operation conditions, and this is called burn-in. The idea of burn-in is to eliminate the probability of infant mortality failures and to screen a more robust population of capacitors that have to operate under extreme conditions. These applications are typically space and military systems. In this study, we will report on the perturbation of the burn-in to the overall reliability of the multilayer ceramic capacitors. This will be analyzed both statistically and also with a physical model. Furthermore, we will speculate on going beyond meantime to failure (MTTF) type analysis in screening pristine populations of capacitors to consider a more robust route to failure analysis in MLCCs.

Determining the Effect of Burn-in Process on Reliability of X7R Multilayer Ceramic Capacitors

Base metal electrode (BME) multilayer ceramic capacitors (MLCC) are widely used in aerospace, medical, military, and communication applications, requiring a high level of reliability. Thus, determining the lifetime reliability of MLCCs is critical to provide more reliable components and no weak links to the electrified infrastructure. The burn-in test is a screening procedure used to eliminate components with a higher probability of infant mortality failures. During this process, components are exposed to high temperatures and voltages relative to their design. The effect of the burn-in test on the dynamics of oxygen vacancies was studied using thermally stimulated depolarization current (TSDC), impedance spectroscopy, and in-situ high-accelerated lifetime testing (HALT). The TSDC results unveiled that the burn-in test caused intragranular and transgranular migration of oxygen vacancies, which will not be relaxed afterward. Mean time to failure (MTTF) data gathered from in-situ HALT demonstrated that burn-in tests were not only ineffective at detecting infant mortality failures but also had a negative impact on the reliability of BME MLCCs. The tracking of MTTF with various models, as well as future challenges to shape changes in the population distribution around these MTTF differences with variables such as electric field, temperature, and time will be discussed.

Electrical Characterization of Vanadium and Yttrium Co-Doped BaTiO3

Co-doping strategies for BaTiO3 have been investigated to improve the reliability and lifetime of multi-layered ceramic capacitors (MLCC). Miniaturization of electronic devices requires more reliable MLCCs without sacrificing performance. Dielectric breakdown of BaTiO3 in MLCCs has been the primary challenge to overcome in the ceramic community. Co-doping acceptors and donors, such as vanadium and yttrium, have been explored as co-dopants in BaTiO3 to reduce electronic and ionic carriers in BaTiO3 via Fermi level pinning. The multi-valent state and limited solubility of vanadium in BaTiO3 presents a significant challenge to understanding the defect structure in the material. Co-doped samples were prepared using a solution doping method, and the sintered samples were characterized using impedance spectroscopy measurements in air and under controlled atmosphere. The data were fit using both impedance and modulus formalisms against a library of equivalent circuits to determine the best fit with the simplest circuit. XRD and impedance spectroscopy data will be presented to show the solubility limit of the dopants. Additionally, the bulk and grain boundary resistances as a function of temperature and pO2 will be examined to better understand how the dopants may lead to improved electrical degradation resistance. This research was funded by the Air Force Office of Scientific Research under grant FA9550-19-1-0222.

Doping Mechanisms and Electrical Properties of Sodium Niobate

The work to be presented highlights electrical characterisation via impedance spectroscopy and the effects of acceptor and donor doping on the electrical properties, and microstructure of NaNbO3 (NN). As with any perovskite, it should be possible to selectively target the A-site (Na) or the B-site (Nb) by changing the nominal stoichiometry and selecting a suitable dopant ion. Acceptor doping the B-site will generate oxygen vacancies and promote ionic conductivity (Nb5++(5-x)/2 O2- = Mx+) whereas donor doping the A-site or the B-site will eliminate oxygen vacancies (Na+=Mx++(x-1)/2 O2-) or generate metal vacancies (eg xNa+=Mx++(x-1)V, where V is an A-site vacancy). B-site Ga-acceptor doped NN displays an enhancement of oxide-ion conduction compared to undoped NN; and is consistent with the expected defect chemistry. B-site W-donor doped NN displays very low conductivity that is consistent with electronic conduction across a band gap and is also in agreement with the expected defect chemistry of NN. In contrast, nominally B-site Mg-acceptor doped NN demonstrate a significant drop in conductivity which is suggestive of A-site donor doping as opposed to B-site acceptor doping. In all cases, the solid solution limits are low, typically less than 1 at% and the results demonstrate the importance of using electrical characterisation to elucidate the doping mechanisms in NN.

Defect-induced, ferroelectric-like switching and adjustable dielectric tunability in antiferroelectrics

Antiferroelectrics (AFEs), which are characterized by antiparallel cation displacements and a field-induced reversible AFE-to-ferroelectric (FE) transition, exhibit great potential for dielectric, electromechanical, and electrothermal applications. While extensive work has been done to understanding how to modulate the polarization of FEs, considerably less attention has been given to AFEs. Here, we demonstrate that point defects (introduced during the growth process) can strongly influence and effectively regulate the switching behavior of AFEs. In AFE PbZrO3 films grown via pulsed-laser deposition, decreasing oxygen growth pressure (and thus varying the adatom bombardment energy) induces exotic “FE-like” polarization switching even though the films remain the typical AFE orthorhombic phase. This “FE-like” switching arises from bombardment-induced point defect complexes which serve as traps for the domain walls (phase boundaries) and impact the AFE-to-FE and FE-to-AFE transitions. We extract the effective trap-energy values with temperature-dependent switching kinetic characterizations. Finally, we demonstrate adjustable dielectric tunability of the PbZrO3 films by changing the concentration of defect complexes, and the conversion between “positive” and “negative” tunability with field poling of the film, in which defect complexes transiently stabilize the poled FE phase.

0-3 Magnetic Composites for Power Electronics, Sensing, and Radio-Frequency Applications

Magnetic composites with 0-3 connectivity, whereby 0-dimensional magnetic particles are embedded in a 3-dimensional matrix, offer potential in a wide variety of engineering applications. This potential stems from the ability to combine and tune the properties of a composite by controlling the materials and ratios of the various phases that make it up. This work will explore past research on two kinds of 0-3 magnetic composites and their applications in power electronics and radio-frequency (RF) devices. Composites formed from embedding non-conductive iron-oxide nanoparticles in a matrix of nickel or permalloy, which have high magnetic permeabilities, allows for the fabrication of planar inductor and transformer cores that can operate at high frequencies for improved efficiency. Composites formed from hard magnetic barium hexaferrite particles in a polydimethylsiloxane (PDMS) matrix allows for the low-temperature (<200°C) fabrication of self-biased magnets for use in non-reciprocal RF devices such as circulators and isolators. Current work towards 0-3 magnetic composites that can be used in other areas, such as electromagnetic interference (EMI) shielding, may also be shared to further demonstrate the usefulness of these materials in engineering applications.

Investigation of optimal pulsed laser deposition parameters for the growth of single crystal barium hexaferrite thin films

In the growth of single crystal thin films, it is imperative to understand the processing-property relationship between film growth parameters and the resultant film. Barium hexaferrite (BHF) thin films are a pervasive materials system, as BHF thin films exhibit lower magnetic losses and do not require high, applied magnetic fields during high frequency use, unlike their bulk counterparts or magnetic garnets. Optimizing the crystalline quality will decrease intrinsic damping and improve high frequency performance. In this work, barium hexaferrite ceramic powder was first produced via traditional solid-state synthesis methods and pressed into dense pellets. These pellets were then used for film growth by pulsed laser deposition (PLD), and parameters such as deposition temperature, laser energy, and deposition time were varied to optimize film deposition; the parameters ultimately chosen were 850°C, 400 mJ, and 2 hours. Deposited films were characterized via Raman spectroscopy, X-ray photoelectron spectroscopy, X-ray diffraction, and magnetometry measurements. These methods confirmed that the expected hexagonal phase, elemental composition, epitaxial orientation, and magnetic behavior were achieved at the optimized parameters. This work informs future work with other adjacent hexaferrite materials systems and their thin film deposition via PLD.

Magnetism in Epitaxial Uranium Dioxide Thin Films

Actinide materials have various applications that range from nuclear energy to quantum computing. Mostly of current efforts have focused on bulk actinide materials. Tuning functional properties by using strain engineering in epitaxial thin films is largely lacking. In this talk, I will talk about the synthesis of a variety of uranium oxides by pulsed laser deposition. In the second part of my talk, using uranium dioxide (UO2) as a model system, I will talk about strain engineering in actinide epitaxial thin films and investigate the origin of induced ferromagnetism in an antiferromagnet UO2. We found out that UO2+x thin films are hypostoichiometric ($x<0$) with in-plane tensile strain, while they are hyperstoichiometric ($x>0$) with in-plane compressive strain. Different from strain engineering in non-actinide oxide thin films, the epitaxial strain in UO2 has been accommodated by point defects such as vacancies and interstitials due to the low formation energy. Both epitaxial strain and strain relaxation induced point defects such as oxygen/uranium vacancies and oxygen/uranium interstitials could distort magnetic structure and result in magnetic moments. This work reveals the correlation among strain, point defects and ferromagnetism in strain engineered UO2+x thin films and the results offer new opportunities to understand the influence of coupled order parameters on the emergent properties of many other actinide thin films.

Anomaly in Hall resistance: Topological Hall effect or two-channel anomalous Hall effect?

Topological Hall effect (THE) due to stable non-trivial chiral spin configurations such as skyrmions has promising applications in next generation high density nonvolatile memories, energy efficient nanoelectronics, and spintronic devices. Recently, an alternative explanation based on two anomalous Hall effects with opposite polarity due to the existence of two magnetic phases has been proposed to explain the phenomena. Spin chirality depends on competition between exchange interaction, anisotropy and Dzyaloshinskii-Moriya interaction. Therefore, we have systematically investigated the origin and tunability of THE like feature in single layer and bilayer $SrRuO_3$ thin films by considering the above-mentioned factors. Comparison of magnetic property, magnetoresistance and Hall resistivity of these samples reveal that anomaly in Hall resistivity certainly has real space topological effects in it.

Voltage-Controlled Topotactic Phase Transformations in Perovskite Cobaltites for Magneto-Ionics

Magneto-ionics offers potential routes to low-power voltage-based control of magnetism, which is highly attractive for next-generation data processing and storage. Electrolyte gating is a powerful approach to magneto-ionics, where electrolytes such as ionic liquids/gels are employed in transistor-type devices for voltage-control of magnetic properties. In this talk, I will review recent advances in this area from my group, focused on voltage control of perovskite oxides, particularly cobaltites. I will focus on our work on reversible electrochemical toggling between ferromagnetic metallic perovskite and antiferromagnetic insulating brownmillerite phases via ion gel gating. This has been achieved across the entire $La_{1-x}Sr_xCoO_{3-\delta}$ phase diagram, studied via operando synchrotron X-ray diffraction, magnetometry, transport, and optical spectroscopy, including gate voltage hysteresis loops. Such advances enable voltage-based control over magnetic, electronic, optical, and thermal properties over extraordinary ranges, with substantial application potential.

Magnetic garnets for spintronic and photonic devices

Magnetic insulator thin films provide unique functionality in spintronic, magnonic, and magnetooptical devices such as magnetic memory, magnetic logic, and photonic integrated circuits. Iron garnets exemplified by yttrium iron garnet are ferrimagnetic insulators with a wide range of properties manipulated by cation substitution. We use pulsed laser deposition to produce films and multilayers of rare earth and Bi-substituted garnets and garnet/metal heterostructures with tunable magnetic anisotropy, damping, and compensation temperatures. These materials reveal a Dzyaloshinskii-Moriya interaction that yields chiral domain walls, and high speed domain wall motion driven by spin orbit torques or by spin waves. Iron garnets also exhibit magnetooptical activity and high transparency in the infrared, and we show how garnets grown on silicon can be used in integrated magnetooptical isolators to control the flow of light in photonic circuits. We will focus on the role of cation site occupancy in determining the properties of garnets and other complex oxides, and we will describe a new class of multiferroic perovskites where ferroelectricity is introduced using antisite defects.

Active Tayloring of Funtional Oxides to Couple Electronics to Spintronics

Spintronics have been proposed as a next-generation, low energy, high speed computational and data storage paradigm. However, after a half century of development, silicon based electronic technologies are exceeding fast and entrenched in virtually every consumer technology - everything from computers and phones to flashlights and pens. Successfully integrating spintronics with existing electronic technologies will allowing the advantages of each to be leveraged to achieve superior technologies. Magneto-ionic control of materials is one approach to achieve this integration. In these magneto-ionic systems, a functional ionic material (usually oxide, nitride or hydride) or ionic material/functional metal heterostructure, is subjected to a large electric field, exerting a force on the ions, facilitating their re-arrangement. These re-arrangements transform the magnetic properties of the ionic material or metallic underlayer. This talk will discuss a handful of systems in which magneto-ionics was used to control magnetic ordering and behavior in (La, Sr)(Co, Mn)O3, GdO/Co, and GdFe/NiCoO systems. In each case, magneto-ionics manifests in unique ways, highlighting the challenges of designing systems in this field. We will also highlight other recent works, emphasizing the use of other ions (hydorgen, nitrogen), the current challenges in the field (speed and temperature), and efforts to overcomes these issues.

Polarization dynamics during multiferroic oxide thin film growth

Multiferroic oxide films with a thickness of just a few atoms can now be grown with a precision matching that of semiconductors. This opens a whole world of functional device concepts and fascinating phenomena that would not occur in the expanded bulk crystal. The successful integration of multiferroics, with coexisting electric and magnetic long range orders, into nanoscale devices relies however on our ability to engineer deterministically polarization states in the application relevant ultrathin regime. Here I will show how nonlinear optics, combined with conventional reflection high energy electron diffraction monitoring tools, enables the in-situ investigation of the ferroelectric polarization during the thin film epitaxial growth. Beyond the unprecedented access to the emergence of ferroelectricity and domain formation during the epitaxial deposition, we investigate transient polarization states originating from the evolving charge-screening environment of the oxide thin film growth process. We show the impact of lattice and interface chemistry as well as depolarizing field-related effects in-situ and identify routes towards the establishment of robust polarization states in the ultrathin regime. Finally, I will show how we use polarization dynamics insights to engineer macroscopic ferroelectric responses and beyond-binary-behavior for emerging applications.

Insights on Structure-Property Correlations in Memristive Devices

Hafnia based resistive random-access memory (RRAM) devices are promising candidates as next generation non-volatile memories and are appealing because of the compatibility with back-end-of-line processes in the current semiconductor fabrication process. Thus, improving device reliability is crucial for their use in future memory applications. Understanding the influence of the microstructure and the atomistic processes involved will help enhance device performance. As shown in our recent work, grain boundary engineering the hafnia layer in TiN/HfO2/Pt RRAM can yield forming free devices with threading grain boundaries that act as preferential pathways for conductive filament formation. The techniques that will be presented provide examples of how to correlate the local structure to the device properties. In situ TEM experiments that involve heating as well as biasing have been carried out on devices mounted on MEMS-based chips by FIB in situ lift-out techniques and are key towards providing a component specific understanding of the physical/chemical mechanisms involved. Furthermore, we have performed a temperature-dependent study to directly monitor the microstructure evolution of the hafnia layer directly inside the microscope, thus giving a unique insight into the mechanism and formation of grain boundaries.

In-situ Characterization of Nanomaterials Using Advanced Electron Microscopy

In this presentation, several of our research examples are selected to highlight the in-situ electron microscopy characterization of strongly correlated electronic materials at atomic and nanometer scales. I will focus on the correlation of electrons, spin and lattice and their degrees of freedom as well as the critical roles of heterogeneity, interfaces, and disorder under external electric, magnetic and photonic excitations to reveal materials' response. For instance, I will show 1) how a new two-color scheme based on photoinduced near-field electron microscopy (PINEM), one for temporal gating, the other for pumping to achieve nm-fs spatiotemporal resolution to visualize ultrafast dielectric response and insulator-to-metal transition in VO2 nanowires; 2) how atomically resolved imaging and electron energy-loss spectroscopy at 10K can be used to understand the interface-enhanced superconductivity in mono-, and a-few-atomic-layer FeSe on SrTiO3; and 3) how to use Lorentz phase microscopy to explore the intriguing transformations among various topological chiral spin states including skyrmions under applied magnetic field at various cryogenic temperatures. The author would like to acknowledge the electron microscopy group at BNL for assistance. The work was supported by the US DOE/BES-MSD under Contract DE-SC0012704.

Revealing strain evolution in catalytic ceria nanoparticles via in-situ Bragg Coherent Diffraction Imaging

Ceria is an important catalytic material with applications in automotive three-way catalysis, solar-driven water splitting, and solid oxide fuel cells. Recent studies have shown significant reduction at ceria's surface, which may enhance catalytic activity, but the extent of the reduction away from the surface and the corresponding lattice structure change remain unknown. To access this information, we use Bragg coherent diffraction imaging to extract the full 3D strain distribution inside ceria nanoparticles, as it evolves during in-situ chemical cycling. We find that the surfaces of the nanoparticles expand in reducing environments, and contract in oxidizing environments, consistent with oxygen vacancies forming and filling, respectively. The strain induced by the vacancies extends tens of nanometers into the nanoparticle, indicating changes well below the surface. Such in-situ studies, enabled by coherent x-ray imaging, will guide the design of optimal crystallite morphologies for catalytic applications. Work supported by the US Department of Energy, Office of Science, Basic Energy Sciences, Materials Science and Engineering Division and the US National Science Foundation (DMR 2130831). Work at the 34-ID-C beamline of the Advanced Photon Source supported by the US Department of Energy, Office of Science, Basic Energy Sciences, under Contract No. DE- AC02-06CH11357.

High Temperature X-Ray Diffraction of Sputtered Thermite Reactive Nanolaminates

Nanoenergetic materials with highly tunable ignition and actuation have applications in military and commercial sectors. Thermite is an inorganic energetic of interest due to the highly exothermic reduction-oxidation reaction that occurs between metal and oxide constituents, resulting in self-sustaining heat production. If the high energy release and tunability provided by the diverse chemistries of inorganic energetics is to be utilized, a fundamental understanding of the initiation and propagation processes in nanoenergetic materials such as thermite is necessary. In-situ high temperature x-ray diffraction (HTXRD) studies on thermite provides opportunities to better understand chemical reactions between fuel and oxidizer as the reaction progresses. This project discusses HTXRD of multilayered, thermite reactive nanolaminates. To probe for potential reaction intermediates, CuO/Mg and CuO/Zr thermite films are ramped at 2 °C/min to a final temperature of 900°C in an inert environment. Results indicate that CuO/Mg RNLs exhibit transient eutectic liquid formation during reaction, with mixed CuxMgy intermetallics temporarily appearing ~565°C. In contrast, Zr/CuO RNLs demonstrate O2 dissolution in Zr ~285°C followed by the formation of ZrO2 ~515°C. These findings provide insight into the mechanisms of energy transfer in thermite RNLs, creating potential for highly tunable, reliable energetic materials.

Surface-Oxide-Driven Evolution of the Anomalous Hall Effect in MnBi2Te4 Thin Films: Implications for the Quantum Anomalous Hall Effect

Understanding interfacial modification of functional properties of magnetic topological insulator thin films is crucial for designing and manipulating properties and ultimately technological applications. This is especially critical for A-type antiferromagnets such as MnBi2Te4 where the magnetic compensation is dependent upon the thickness. In this talk I will discuss the large electronic and magnetic response that is induced in MnBi2Te4 by controlling the propagation of surface oxidation. This oxide is strongly confined to the top layer yet drives a dramatic reversal of the sign of the anomalous Hall effect driven by finite thickness magnetism, which indicates that the film splits into distinct magnetic layers each with a unique electronic signature. This suggests that magnetic perturbations resulting from surface oxidation may play a non-trivial role in the stabilization of the quantum anomalous Hall effect in this system and that understanding targeted modifications to the surface may open new routes for engineering novel topological and magnetic responses in this fascinating material.

New Applications of Direct Detection, Electron Counting Cameras for Materials Science

Direct detection cameras are changing the transmission electron microscopy landscape by delivering outstanding data quality at extremely low electron dose rates. These cameras have supercharged the cryo-EM research community and are increasingly valuable to materials scientists as research interests continually extend to more beam-sensitive materials. This presentation will review the fundamentals of direct detection sensor design, concepts of electron counting, and how these technological advancements enable low dose TEM imaging, diffraction, and in-situ experiments. Key results from a range of materials applications will be highlighted, 2-D materials, perovskites, and other beam sensitive samples. Extending direct detection technology to low-kV TEM analysis, 4D STEM, EELS, and EBSD will also be discussed.

Operando two-terminal devices inside a transmission electron microscope

Advanced nanoelectronics materials are guided by their electric performance under operative conditions. A detailed understanding of the correlation between the electrical properties and the material structure at the nano and sub-nano meter scale is nowadays crucial in electronic industries. In this context, electrical biasing in situ transmission electron microscopy (TEM) using MEMS-based platforms is a promising technique for nano-characterization under operative conditions. However, electrical biasing experiments using MEMS-based platforms have been to date overshadowed due to stray leakage current paths formed during focused ion beam (FIB)-based sample preparation routines, leading to over-estimated electric responses and therefore, ambiguous in situ observations. Here, we undertook this problem by establishing a short-circuit-free FIB methodology that shows an improvement of at least five orders of magnitude in leakage current in comparison with conventional-FIB approaches and the reported values found in the literature. Our methodology enables current measurements as low as 10 pA, allowing realistic and reproducible biasing operations of various types of oxide-based devices. In addition, to further demonstrate device connectivity inside a TEM we performed electron beam induced current (EBIC) experiments.

In-situ characterization of perovskite oxides with neutron reflectivity

Hydrogen induced topotactic phase transitions have recently become an area of intense study with the emergence of superconductivity in reduced nickelate perovskite oxides and new memristor and neuromorphic-based technologies. Changes in oxygenation and hydrogenation produce electronic structure changes that manifest broadly as metal-insulator transitions, emergence of superconductivity, or magnetic ordering. However, poor understanding of the fundamental mechanisms of hydrogenation and a lack of systematic studies makes the understanding of these emergent properties difficult to digest. The first step toward understanding hydrogen induced transformations is the quantification of the amount and the distribution of hydrogen vs. formation of oxygen vacancies throughout the lattice during topotactic phase transitions. Using La0.7Sr0.3MnO3 as a prototypical example, we report on the mechanism for oxygen vacancy formation and ordering by hydrogenation. During hydrogenation, the ordering of initially randomly distributed oxygen vacancies produces a perovskite to brownmillerite phase transition. In-situ NR measurements reveal that the phase transformation of LSMO is driven by a transient mechanism limited to the film surface without protons ever entering the lattice. These results demonstrate the strength of in-situ studies in understanding phase transformations and emergent properties in oxide systems.

Probing Structure Evolution in Functional Oxides by in-situ Neutron Diffraction

Understanding dynamic structure changes in materials is key to probing the mechanisms of materials synthesis and process, and the origin of materials functional behaviors. However, post-experiment characterizations can hardly capture transients during a continuous evolution. Thanks to the high-flux and deep-penetrating neutrons at the Spallation Neutron Source, the time-of-flight neutron diffraction is enabled to characterize the structure evolution in atomic scales in real time, when the equipped sample environment mimics the material processing. The neutron diffraction has a unique capability to resolve fine structures, by detecting key lightweight elements and differentiating neighboring transition-metal-elements. Taking the high-voltage $LiNi_{0.5}Mn_{1.5}O_4$ spinel cathode for Li-ion battery as an example, in-situ neutron diffraction probes the kinetics of the atoms interdiffusion, ordering, and the crystal structure transition under annealing at different atmospheres. The structure destabilization mechanism is correlated to the Mn^{4+} reduction and the Mn^{3+} migration. Additional examples of shape-memory oxides and ferroelectric oxides are also given. The deformation mechanisms under stresses are revealed, such as anisotropic lattice straining, ferro-domain switching, and displacive phase transformation, which determine the materials functional properties.

Thermal Behavior Transition of Graphite under High Pressure

Carbon can exist in many phases in nature, including some extreme cases such as graphite, where single atomic layers bonded together via Van der Waals force, cubic diamond with highest thermal conductivity and hardest known material, and amorphous carbon without any crystal structure. Thermal properties exhibited by these different phases are drastically distinct. Transition between different phases can be induced under certain temperature and pressure conditions. For example, transition from graphite to diamond; C60 to amorphous carbon, and others. Structural changes during these transitions are studied rigorously with simulations and probed indirectly with XRD. One big element missing is how these transitions would affect the heat conduction ability. In this work, we demonstrate the thermal properties evolution when graphite experiences phase change under high pressure environment created in Diamond Anvil Cell. We found that phase transition can affect thermal conduction from several perspectives, including: change of bond type (from sp2 to sp3), bond distortion, and grain boundary formation, which gives a rather complicated trend in thermal conductivity. This work paves a way to explore the phase diagram of graphitic materials from thermal perspective, may intrigue interests and efforts to explore and synthesize graphitic materials with transition phases with desired thermal properties.

Ferroelectric Doped Bismuth Ferrite for Photocatalysis

$BiFeO_3$ (BFO) is a type-I multiferroic with noteworthy ferroelectric behaviour at room temperature. Scaling the BFO particle size down to the nanometer increases the total magnetization. The large surface to volume ratio is useful in photocatalysis, photovoltaics, and electrocatalysis. We utilize multivalent codoped BFO nanoparticles (NPs) for photocatalysis exploiting their narrow electronic bandgap. $BiFe_{0.95}Mn_{0.05}O_3$ (BFM) NPs with co-incorporated mono-, di-, or tri-valent cations (Ag^+, Ca^{2+}, Dy^{3+}; MDT) show an increased photocatalytic performance under ultrasonication and in acidic conditions. Piezoresponse of these NPs becomes a vital factor for photocatalysis under ultrasound. We relate the photocatalytic activity of the doped NPs with their structural, optical, and electrical properties. Monovalent doping enhances the photocatalytic activity of these NPs approximately three times. The increase of spontaneous polarization by the mono- and tri-valent doping enhances the photocatalytic performance along with stronger light absorption in the visible range, low recombination rate of charge carriers, and larger surface area of the NPs. Divalent (Ca^{2+}) doping reduces the polarization, whereas trivalent (Dy^{3+}) and monovalent (Ag^+) cations provide an increase of polarization. The depolarization field in these single domain ferroelectric photocatalysts reduces the recombination of the photoinduced charge carriers.

Operando measurement on lithium-ion batteries using a laboratory X-ray diffractometer

Structural characterization of cathode materials on lithium-ion batteries (LIBs) is a key to understand its electrical stabilities such as endurance and degradation. X-ray diffraction (XRD) provides the crystal structure of the materials including constituent atoms, atomic sites occupation, bond length and lattice parameters. However, most of the XRDs used in the laboratory equip with copper X-ray source which emits about 8 keV energy of X-rays. The cathode materials used in the LIBs typically include 3d orbital transition metals, e.g. Co, Mn, Ni and Fe, and those elements are highly absorbing for 8 keV X-rays and generate X-ray fluorescence. Due to that, diffraction peak signals are generally weak, and background is high by the excited XRF. Considering that the X-rays penetrate deeper on transmission geometry than on reflection geometry, it is essential to have high energy X-rays in the transmission geometry to probe the materials of interest. In this paper, we will discuss operando XRD measurement on a laboratory XRD that equips with a high-flux and high-energy X-ray source and a newly developed semiconductor based 2-D detector that stops the high-energy X-rays, applied on a pouch cell structured $LiFePO4$ based LIB. We were able to identify its ion conduction path and cycle rate dependence of the charge-discharge property through the experiments.

AXON: Experiment-to-Analysis Software for High Performance Transmission Electron Microscopy Studies

State-of-the-art in-situ systems enable the transmission electron microscope (TEM) to function as a nanoscale laboratory. Researchers can observe a sample's real-time behavior under a variety of environmental conditions and simultaneously deliver and/or source stimuli, such as electrical impulses or temperature changes. Thus, a range of dynamic processes, such morphological changes induced by corrosion, structural changes experienced by a single catalysis particle during its work cycle, or dendrimer formation on a battery surface can be scrutinized. The resulting data sets are often cumbersome to analyze due to their large size and the multiple streams of metadata generated. Protochips recently introduced AXON, a software solution that addresses the pain-points inherent to in-situ studies. Consisting of both a data-collection module (AXON Synchronicity - which connects, stabilizes, and integrates data streams from the microscope, detectors, and in-situ devices) and a free-to-use post-processing analysis software (AXON Studio - that streamlines image sharing, analysis, and trend discovery) the AXON platform reduces the learning curve that has plagued adoption of in-situ TEM techniques and simplifies collaboration among colleagues, enabling identical datasets to be analyzed independently, without the loss of precious metadata associated with the individual images.

Thermoelectric performance and high-field behavior of Pb-free PTCRs

Positive Temperature Coefficient Resistors (PTCR) are used as passive elements in many applications such as heaters, inrush current limiters and motorstarters. To meet the requirements for this variety of different applications, compositional modifications are necessary to yield the desired electric properties, like resistivity, Curie temperature or high-temperature behavior. Furthermore, reliable PTCR ceramics must withstand high voltages and maintain their properties even after high-field and temperature cycling. Common PTCR ceramics consist of BaTiO3. Various A-site substitutes are added to achieve the desired properties. Pb was demonstrated to fulfill the mentioned electric conditions best. However, according to the RoHS directive enacted by the European Union, Pb is considered as a harmful substance which makes the development and investigation of Pb-free alternatives necessary. This study focuses on investigations of the BaTiO3 system and the tuneability of the electrical properties via acceptor and donor doping as well as by substituting Ba by with various isovalent A-site elements. The microstructure is investigated in detail using electron microscopy and X-ray diffraction and correlated to electrical parameters. The gained structure-property relationship serves as crucial input for the development of novel Pb-free protection devices.

Relating the initial microstructural state's grain boundary properties to the abnormal grain growth behavior in SrTiO3

Microstructural control provides an opportunity for controlling the bulk performance of the material. However, efficient microstructural design is still out of reach due to the poor understanding of grain growth phenomena such as abnormal grain growth (AGG), which occurs uncontrollably. Mainly, it is unclear how an initial state of a microstructure in terms of the macroscopic grain boundary character (GBC) and grain boundary energy (GBE) should be to promote or prevent AGG. To investigate the influence of grain boundary energy, SrTiO3 - a material known to exhibit AGG at 1425oC and anisotropic grain boundary energy – is prepared with different low-temperature heat treatment conditions. These processing conditions change the initial grain boundary plane and energy distribution and, thus, influence subsequent grain growth behavior. A grain growth study at higher temperatures is then conducted on the samples to assess for variations in abnormal grain growth behavior. Microscopy techniques will be used to characterize the differences in GB properties between the different initial states. The results from this study will provide an understanding of how initial GB properties relate to abnormal grain growth in SrTiO3.

Understanding Metal-Insulator Transitions in Heteroanionic TiOF with ordered Heteroleptic Polyhedra

Temperature-driven metal insulator transitions(MITs) hold great potential for applications in next-generation electronic devices.Rutile oxides,in particular VO2 and NbO2,have been intensely pursued and are exemplary examples of compositionally simple materials with a single anion exhibiting MITs.The underlying interactions governing the MITs in these oxides arise from a combination of factors:electronic correlations,metal-metal bonding and lattice distortions cooperatively interact to drive the transition.We investigate the behavior of the rutile-structured heteroanionic oxyfluoride TiOF using first principles simulations as a route to disentangle various contributions to MITs in the d1 rutile family.We model the structure of TiOF using edge-sharing fac ordered [TiO3F3] heteroleptic units and explore the effects of magnetism and electron correlation. We find that in the absence of onsite Coulomb interactions (U=0 eV),a metal-metal transition occurs accompanied by a change in crystallographic symmetry with Ti-Ti dimerization.For U> 0 eV,we find the structural transition is disfavored and a ferromagnetic insulating state is dynamically stable.We conclude by contrasting our results with related d1 rutile compounds exhibiting MITs and explain how future work will include effects on the electronic structure from variations in long-range ordering of the fac ordered heteroleptic units.

Novel Integration of Graphene-based Ceramics for Enhanced Supercapacitor Energy Storage

While the conventional active material for high power density supercapacitors, activated carbon (AC), is commonly selected for its high specific surface area, commercial availability, and low cost, it still suffers from issues of electrical conductivity and ageing. Efforts to mitigate these problems, such as creating in situ AC composites with performance-enhancing materials, are successful but often complicate the overall materials synthesis process. In this work, we investigate the efficacy of creating AC composite electrodes using a simple mixing strategy with popular yet unexplored graphene-like ceramics: graphene oxide (GO) and reduced graphene oxide (rGO). It was found that GO increases the rate resistance of commercial AC electrodes by an average of 35.7 percentage points between current densities of 0.1 and 10 A/g, a result that can be ascribed to the varied functional groups of GO, determined via X-ray photoelectron spectroscopy, that stabilize AC particles. Alternatively, rGO increases capacities across a 10,000 cycles stability test by 12-17 % compared to commercial AC electrodes, a result that can be ascribed to the highly conductive rGO that interfaces with AC surfaces, as observed via field emission scanning electron microscopy. This work further establishes the complementary value of GO and rGO as functional ceramic additives for electrochemical energy storage devices.

Grain Boundary Chemistry and Potentials in Co-doped BaTiO3

With high dielectric permittivity, barium titanate (BaTiO3) is one of the most suitable materials as a dielectric, and commercial multilayered ceramic capacitors (MLCCs) are based on BaTiO3. The demand for miniaturization and high performance of electronic devices requires high volumetric efficiency, high capacitance and high reliability of MLCCs. Numerous studies have investigated the effects of dopants on the dielectric properties of BaTiO3. Previous literature shows that co-doping BaTiO3 with Mn and Y limits both the bulk electronic and ionic charge carrier concentrations over broad pO2 ranges by pinning the Fermi-level near the middle of the bandgap. However, it is generally accepted that grain boundaries (GBs) also affect the dielectric properties and lifetime behavior of polycrystalline BaTiO3. In this work, we study the GB defect chemistry and its contribution to the macroscopic conductivity. BaTiO3 pellets are prepared by conventional solid-state reaction method, and the samples are characterized by impedance spectroscopy and scanning transmission electron microscopy (STEM). Impedance spectroscopy data reveal the relative contributions of the bulk and GB resistances as a function of temperature and provide useful data from which GB space charge potential can be extracted. Additionally, STEM data are presented to analyze the electrostatic potential barrier height at the GBs caused by uncompensated defect states.

Predicting the performance degradation of advanced devices exposed to ionizing radiation

It is crucial to ensure the effectiveness and efficiency of semiconductor devices when they work in harsh environments. In this work we are developing a new novel simulation tool that predicts the impact of ionizing radiation on the performance of a semiconductor device. The approach will couple TCAD capability to predict device performance with detailed radiation models to predict charge carrier and lattice defect production. The tool is implemented using the open-source Multiphysics Object-Oriented Simulation Environment (MOOSE). We determine the performance of the devices by calculating local carrier concentration and local current density vs. DS voltage dynamically over time. The results show the relationship of performance degradation rate and irradiation. The final tool will be open source, have robust quality assurance practices, and be able to be referenced in designing more radiation hard devices.

Modeling ErAs nanoparticles in III-Vs using Density Functional Tight Binding

III-Vs embedded with ErAs nanoparticles enable a wide range of optical, electrical, and thermoelectrical properties and are shown to have an important use in Terahertz (THz) technologies. Computational modeling of the electronic structure of these materials is required to investigate some of these properties. Density functional theory (DFT) has been used to model ErAs-GaAs nanocomposite systems. However, the use of DFT is limited by its large computational cost and ability to only investigate small-scale systems. Density functional tight binding (DFTB) offers an approximated DFT with the ability to compute large-scale systems at a much lower computational cost. We investigated the ability to model an ErAs-GaAs nanocomposite structure with DFTB that had been parameterized off its subcomponents, ErAs and GaAs. Parametrization was verified by comparing the band gap and band shape to prior DFT modeling. The results reveal that DFTB can be used to model the nanocomposite system to an acceptable level of accuracy. This new modeling method provides a promising route to investigate larger-scale and more realistic-size composite systems, including semiconductors/semimetals with rare-earth elements.

Phase Control of Two-Dimensional Tin Sulfide Compounds Deposited via Atomic Layer Deposition

Two-dimensional (2D) semiconductor materials are emerging as potential candidates for emerging electronic device applications due to their excellent electrical properties and unique properties such as flexibility and transparency. Among the various potential 2D semiconductor materials, tin sulfide compounds can form polymorphs of different 2D crystals: 2D layered-SnS2 with hexagonal symmetry and 2D layered SnS with orthorhombic symmetry. In addition, since they each have different electrical and optical properties, their controlled growth is important for appropriate target applications. Therefore, we investigated the phase control of tin sulfide compounds deposited via atomic layer deposition according to their deposition temperature and post-annealing conditions. A strategy for controlling the growth of SnS2 or SnS was proposed by investigating phase transition tendencies according to deposition temperature and annealing conditions (temperature, pressure, atmosphere). As a result, it was possible to selectively deposit SnS2 and SnS thin films based on a phase transition study; it was confirmed that SnS2 and SnS thin films exhibited n-type and p-type semiconductor characteristics, respectively.

Effects of Y Doping on Ferroelectric and Electrical Properties of As-deposited $Hf_{1-x}Zr_xO_2$ Thin Films via Atomic Layer Deposition

$Hf_{1-x}Zr_xO_2$ (HZO) thin films are one of the most attractive HfO_2-based ferroelectric thin films for potential applications in next-generation ferroelectric devices. As previously reported, the non-centrosymmetric polar orthorhombic phase (o-phase, $Pca2_1$) which is attributed to ferroelectricity of HZO thin films can be stabilized by doping with Group III dopants. Until now, several Group III elements doped HZO thin films have been fabricated by atomic layer deposition (ALD), but a study on Y-HZO thin films via ALD has not been conducted. In this study, we investigated the effects of Y doping cycles on ferroelectric and additional electrical properties of as-deposited HZO thin films with various compositions via ALD. Through this study, the ferroelectric transition behavior according to the Y doping cycle of as-deposited HZO thin films with various compositions was confirmed for the first time. Especially, in the case of Hf-rich Y-HZO, the $2P_r$ value increased by 3.30 times, up to Y-doping cycles of 4. In addition, it showed the complex advantages of improving the dielectric constant, leakage current density, endurance, and decreasing the coercive field value. This work can contribute to the application of these Y-HZO ferroelectric thin films into semiconductor devices requiring a low thermal budget and excellent electrical properties.

Rutile-TiO_2 thin films deposited by Atomic Layer Deposition using thin SnO_2 seed layer and Sn doping for DRAM capacitor

As dynamic random-access memory (DRAM) devices have been continuously required scaling down, research on high-k materials has been considerably conducted. Among them, rutile-TiO_2 has a higher dielectric constant (90~170) than the other high-k materials. However, since the crystallization temperature of rutile TiO_2 is 750°C, which is unacceptable in the DRAM process, a functional electrode such as RuO_2 has been used. Unfortunately, RuO_2 is reduced to Ru during forming gas atmosphere annealing in a subsequent process. In contrast, TiN used as the bottom electrode of conventional DRAM capacitors is stable and compatible between deposition and other process steps. In addition, SnO_2 has a thermodynamically stable rutile phase at room temperature and can induce rutile-TiO_2 because of negligible lattice mismatch with rutile-TiO_2. Therefore, we investigated the effect of SnO_2-layer insertion and Sn-doping into TiO_2 thin films based with TiN electrodes on structural and electrical properties for use as high-k materials. Rutile-TiO_2 was induced even with a small amount of Sn doping concentration, and it was confirmed an ultra-thin 1 nm SnO_2 seed-layer was sufficient to form rutile-TiO_2. Sn-introduced TiO_2 thin films have improved electrical properties, which suggest the possibility that these devices can be applied as a dielectric for next-generation DRAM capacitors.

Processing Optimization of Novel High Tc Piezoelectric Ceramic

Our recent work has found a novel lead reduced piezoelectric system with enhanced properties at the morphotropic phase boundary (MPB) and a high Curie temperature (Tc = 650 °C). This system exists as a solid solution between a lead-free perovskite, $Bi(Fe_{2/3}Mg_{3/8}Ti_{3/8})O_3$, and $PbTiO_3$. Of particular significance, the compositions within the MPB have a high piezoelectric response (d_{33} = 100 pC/N) compared to previous conventional high-Tc piezoelectric ceramics. This solid solution was prepared via conventional solid-state technique. Through exploration of synthetic optimization, we have learned that processing plays a significant role in electromechanical properties of this material. This work investigates the effect of ball mill time and sintering method (1-step, 2-step and quench) on the piezoelectric properties in this novel high Tc ceramic material.

The Role of Stoichiometry in Mn1-xZnx Ferrite Microwave Absorbers

As 5G technology moves higher into the microwave frequency regime, coupling EM energy into susceptible electronic circuits grows. Minimizing this interference is driving the need for compact, lightweight broadband absorbers to protect sensitive electronics. Manganese zinc ferrite (Mn1-xZnxFe2O4) microparticles have been demonstrated to be magnetically lossy up to 10 GHz and have the potential to meet the energy density and attenuation needs. Optimizing the shielding effectiveness requires improved understanding Mn1-xZnx stoichiometry and the dominant dielectric/magnetic absorption mechanisms for each stoichiometry. We discusses results, to 20 GHz, from experimentally investigating the shielding effectiveness and quality factor attenuation of 30 micron Mn1-xZnxFe2O4 with x=0 to 1 in intervals of 0.1. Complex permeability and permittivity measurements were conducted to identify the absorption mechanism. An increase in absorption was observed in Mn1-xZnxFe2O4 for x=0.6, 0.7. A likely explanation of the observed increase is provided. This work was supported by the Laboratory Directed Research and Development program at Sandia National Laboratories, a multimission laboratory managed and operated by National Technology and Engineering Solutions of Sandia LLC, a wholly owned subsidiary of Honeywell International Inc. for the U.S. Department of Energy's National Nuclear Security Administration under contract DE-NA0003525.

Flash Sinering of BaTiO3

Flash sintering is a fairly novel ceramics densification process, which has gained interest due to its rapid densification effect and the associated possibilities of energy efficiency improvements in ceramic processing and microstructure control. While its general effectiveness has been demonstrated in a range of oxide materials, upscaling remains a major challenge due to the inhomogeneous densification and hot-spot formation. We report on in-situ IR thermometry measurements of flash sintering events in barium titanate green bodies of different geometries. Both the change in temperature during the flash event as well as the temperature gradients on the sample surface were monitored. An expected exponential increase in temperature, proportional to the increase in current density for the flash event, was recorded. However, no correlation between the peak temperature and the sample density could be established. A temperature gradient of up to 200 °C was identified on both sides of the sample in field direction due to heat transport through the electrodes. For non-symmetrical samples the formation of a significant gradient orthogonal to the field direction was recorded. Incubation of the hot-spots due to the differences in heat energy losses through radiation and convection is demonstrated.

Ferroelectric properties measurements of K1-x NaxNbO3 thin films synthesized through aqueous solution deposition

With widespread restrictions and environmental issues related to lead containing compounds, potassium sodium niobate - K1-x NaxNbO3 (KNN) is a promising lead-free piezoelectric material. Although functionality and properties of the material have been well studied in the bulk form of ceramics, comparatively little is known about how it behaves as a thin film. Generally, chemical solution deposition (CSD) techniques reported for fabricating KNN films require the use of hazardous organic solvents (namely, 2-methoxyethanol). In addition, challenges associated with fabricating films at a high quality and with high throughput is a contributor to the current limited applications of KNN in industrial scales. The first logical steps in developing KNN devices is improving the quality of thin films, demonstrating control of their functional properties, and understanding how the atomic structure of the film drives the physical properties. Piezoelectricity properties and structure of KNN films was studied by using high resolution x-ray diffraction (HRXRD), Scanning Electron Microscope (SEM), Atomic-force microscopy (AFM), Transmission electron microscopes (TEM) and measuring Polarization–electric-field (P-E) hysteresis loop. The results described are indicative that the new aqueous deposition technique KNN products can be optimized to compete with those of PZT.

Radio-imaging of electrical discharge in ceramic components

We introduce a novel diagnostic for locating electrical breakdown and its precursors (i.e., corona). Identifying the potential for electrical breakdown is critical to product realization of electrical components and systems. Existing diagnostics indicate electrical breakdown without resolving failure locations (e.g., current/voltage/chemical measurements), locate external breakdown (e.g., optical/UV photography), or locate both external/internal breakdown but with low fidelity (e.g., UHF/acoustic triangulation). To address these technical challenges, we adapt a radiofrequency technique for imaging lightning to the smaller spatio-temporal scales of electrical devices. Using 3-GHz receivers, we demostrate: 1) the ability to image breakdown location(s) at sub-mm and ns scales, and 2) the ability to identify the different breakdown regimes (corona/streamer/arc) based solely on the receiver signals. Supported by the Laboratory Directed Research and Development program at Sandia National Laboratories, a multi-mission laboratory managed and operated by NTESS LLC, a wholly owned subsidiary of Honeywell International Inc. for the U.S. Department of Energy's NNSA under contract DE-NA0003525. This paper describes objective technical results and analysis. Any subjective views or opinions that might be expressed in the paper do not necessarily represent the views of the U.S. Department of Energy or the United States Government.

Dielectric properties of co-doped BaTiO3 relaxor ceramics

The growing demand for 5G mobile products requires higher capacitance devices at a reduced size. As modern fabrication techniques approach a size limit for multilayer ceramic capacitors (MLCC), materials with larger permittivities will be required. Barium titanate, an extensively studied and commercially ubiquitous dielectric material, is a promising candidate as it is commonly used in modern capacitor fabrication, and its properties can be manipulated by introducing rare earth co-dopants. Simultaneous co-doping on both the A and B sites in barium titanate has been shown to depress the phase transition temperature and to induce a more diffuse structural transition. This study was focused on the effects of co-doping barium titanate with rare earth elements with the goal of preparing a dielectric material with a diffuse phase transition to maintain high capacitance over a broad temperature range, with a K of at least 10,000 at 1 kHz, a low temperature coefficient, and a low dielectric loss. Samples in the shape of ceramic discs were synthesized using standard solid state processing techniques. Characterization techniques used in this study include x-ray diffraction, density measurements, impedance spectroscopy, SEM imaging, diffuse reflectance, and measurement of dielectric properties.

Measurement of Thermal Conductivity of Highly Anisotropic 2D Polymerized Carbon

2D polymerized fullerene sheets with weak van der Waals interfaces offer many advantages as future optoelectronic materials. To create the carbon allotrope studied in the work, magnesium is used as a dopant to create covalent bonds between fullerenes within a single plane. By locking rotation between fullerenes in the same sheet and losing a degree of motion, heat transfer is enhanced within the polymerized sheets, but remains low between the van der Waals bonded layers. This crystal structure remains even after the magnesium is deintercalated by a dilute acid, and the covalent bonds are preserved. In this work, we measure the thermal conductivity of magnesium doped fullerenes using time-domain thermoreflectance (TDTR), an optical pump-probe technique for measuring thermal properties of thin films and nanostructures. We also present spatially resolved thermoreflectance maps of Mg4C60 and the resulting deintercalated C60. Molecular dynamics simulations demonstrate extreme anisotropy in both the magnesium doped and deintercalated samples, and predict an average thermal conductivity significantly higher than that of a single-crystal fullerene, in agreement with experimental results.

Applicability of Aerosol Deposition Process in Preparation of Ceramic Coatings

The Aerosol deposition (AD) is a novel coating deposition process, which allows the fabrication of dense, nanocrystalline ceramic coatings via a room temperature impact consolidation mechanism. Without the necessity of heat treatment, it is capable to retain the original properties of ceramics and metals, and prohibit interfacial material reactions. Compared with other conventional methods, e.g., PVD, CVD and thermal spray, the AD has several advantages, including low temperature deposition process, enabling integration of ceramic coatings with a wide range of metallic and polymeric substrates, flexibility in coating composition, and relatively high deposition rates. In this research, α-Al2O3 powders with low relative permittivity (9.5), high thermal conductivity (26 W*m-1K-1), high electrical resistivity (above 1014-1016 Ω*m) and adequate breakdown strength (~30 kV*mm-1) were successfully deposited on different substrates (copper, glass, polymers, etc.) for insulating applications. The spraying process and particle impact mechanism would also be observed and analyzed by both numerical and experimental studies.

Bonding and Crystal Structure Properties of YB_{66} at High Pressure Conditions

YB_{66} is a technologically important material and has a unique icosahedral structure (Fm3c (No. 226)) [1,2]. Among all yttrium borides, YB_{66} is a hard material and exhibits Knoop hardness of 26 GPa and finds importance as a potential thermoelectric material [3]. Furthermore, YB_{66} is a soft X-ray monochromator material used for dispersing synchrotron radiation in the energy range from 1 to 2 keV [4]. We have systematically investigated the bonding and crystal structure properties through B-edge X-ray Raman spectroscopy and high-pressure x-ray diffraction up to 50 GPa. A gradual change in the σ and π bonding features are observed under compression. We will discuss both the bonding and crystal structural changes at high pressures in detail. This work was supported by the US Department of Energy/National Nuclear Security Administration through the Chicago/DOE Alliance Center (CDAC), Award DE-NA0003975 and NSF-DMR grant DMR-2119308.

Conduction mechanism of ZrO_2-based nano-laminates structure for suppressing the leakage current

ZrO_2-based thin films have widely used for semiconductor devices such as gate insulater, DRAM capacitors owing to their relatively high bandgap and superior dielectric constant. Meanwhile, as the size of the devices is downscaled, it is important to suppress the leakage current without the degradation of dielectric constant. To improve the leakage current, the research have been conducted to insertion or doping of SiO_2 or Al_2O_3 between ZrO_2 thin films. It is known that SiO_2 or Al_2O_3 layers were effectively reduced the leakage current with sustaining the dielectric constant due to a high bandgap. However, since the leakage current is affected by film quality, grain boundary, or the interface between metal and insulator, it is ambiguous to explain the effect of SiO_2 or Al_2O_3 with only bandgap. In this regard, the conduction mechanism is one of the most method to understand the cause of leakage current. Therefore, we investigated the electrical properties of the MIM capacitors with ZrO_2-based nanolaminates structure. Based on the conduction mechanism and conductive atomic force microscopy (C-AFM), it will be observed the charge transport phenomenon in thin films, and suggest the solution to reduce the leakage current without decreasing the dielectric constant.

Piezoelectric Properties of Gadolinium-Modified Bismuth Titanate Ferrite ($Bi_5Ti_3FeO_{15}$) for High-Temperature Applications

Bismuth titanate ferrite $Bi_5Ti_3FeO_{15}$ (BTF) has recently attracted considerable attention as a typical multiferroic oxide, wherein ferroelectric and magnetic orders coexist. The ferroelectric order of BTF implies its piezoelectricity, because a ferroelectric must be a piezoelectric. However, no extensive investigations have been carried out on the piezoelectric properties of BTF. Considering its high ferroelectric-paraelectric phase transition temperature (T_c ~ 761 oC), it is necessary to analyze the piezoelectricity and thermal stabilities of BTF, a promising high-temperature piezoelectric material. In this work, we report significantly enhanced dc electrical resistivity and piezoelectric properties of BTF through rare-earth element gadolinium (Gd) ions compositional adjustment. The nominal compositions of $Bi_{5-x}Gd_xTi_3FeO_{15}$ (abbreviated as BTF-100xGd) have been fabricated by conventional solid-state reaction method. The composition of BTF-8Gd exhibits a significantly enhanced dc electrical resistivity, and also exhibits enhanced piezoelectric properties with a high piezoelectric coefficient d_{33} of ~24 pC/N and a high T_c of ~784 oC. Moreover, the BTF-8Gd exhibits good thermal stabilities of piezoelectric properties up to 400 oC. These properties demonstrate the great potentials of Gd-modified BTF for high-temperature piezoelectric applications.

Enhanced Energy Storage Properties by Constructing Relaxor Ferroelectrics in Ternary $Bi_{0.5}Na_{0.5}TiO_3$-$Bi_{0.5}K_{0.5}TiO_3$-$Ba(Al_{0.5}Ta_{0.5})O_3$ Ceramics

Dielectric ceramic capacitors are the key electronic elements for pulsed power systems, but low recoverable energy storage density of dielectric ceramic materials limits the miniaturization of pulsed power systems. Perovskite-structure $0.80Bi_{0.5}Na_{0.5}TiO_3$-$0.20Bi_{0.5}K_{0.5}TiO_3$ (BNT-BKT) has attracted considerable attention because of its large saturated polarization (Pm) at the morphotropic phase boundary (MPB) composition, where rhombohedral and tetragonal phases coexist. In this work, ternary BNT-BKT-$Ba(Al_{0.5}Ta_{0.5})O_3$ (BNT-BKT-BAT) ceramics have been investigated to seek dielectric ceramic materials with high energy storage density. The dielectric, ferroelectric, and energy storage properties of BNT-BKT-BAT ceramics are investigated in details. By composition optimizing, high breakdown strength and slim hysteresis loops are simultaneously obtained due to the refined grain size and the formation of polar nanoregions (PNRs), respectively. An excellent recoverable energy storage density Wrec of 3.84 J/cm^3 and a large energy efficiency η of 84% are concurrently achieved. Meanwhile, the BNT-BKT-BAT ceramics show remarkable temperature and frequency stabilities and ultrafast discharge rate, indicating that ternary BNT-BKT-BAT relaxor ferroelectrics have wide application prospects in the field of pulsed power capacitors.

Frequency dependence of domain switching in lead zirconate titanate

Certain characteristics of ferroelectric materials change as a function of frequency of the applied electric field, including the piezoelectric coefficient, coercive field, and domain switching behavior. Poling ferroelectric materials through the application of an electric field aligns domains through domain wall motion. Time-resolved, in situ synchrotron x-ray diffraction patterns were measured on bulk, poled $Pb(Zr_{0.4}Ti_{0.6})O_3$ (PZT) piezoceramics under applied electric fields with frequencies of 1-1000Hz. The fraction of switched domains, η, was determined using the intensities of the 002 and 200 reflections at each time stamp. Domain wall motion is characterized by the time it takes to switch domains, τ, or the time constant. In the present work, we demonstrate a logarithmic frequency dependence of η. The time constant of 6.90 x 10^{-5} was calculated from the 1000Hz experiment which most clearly demonstrated the switching behavior. Switching fractions were also observed in the 1-100Hz data and fit with the 1000Hz time constant model. Thus, a logarithmic frequency dependence of η is demonstrated across four orders of magnitude.

Low-Temperature Oriented Bottom Electrodes on Amorphous Surfaces

Demonstrating ferroelectricity in c-axis oriented wurtzite nitride thin films (Al1-xScxN, Al1-xBxN, and Zn1-xMgxO) has attracted recent attention for application in integrated memory devices because it can be grown at substrate temperatures well below 400 °C and is compatible with back end of the line processing boundary conditions as well as polymeric implementations. However, ferroelectric switching, at this point, requires 0001 textured material, thus new methods are needed to implement this orientation on substrates that are restricted to abide by CMOS ground rules and chemical compatibility with a diverse materials palette that includes organic and hybrid options. This work investigates how amorphous oxides, silicides, nitrides and polymers, in conjunction with FCC and HCP metals, can be used to promote crystallographic texture on a crystallographically agnostic surface. We find that HCP, FCC, and BCC metals can grow with their close-packed planes parallel to the growth surface providing sufficiently well controlled interfaces and landing energies. The high degree of crystallinity will promote c-axis oriented wurtzite nitride thin films at low temperature conditions with possibility for both insulating and conducting stack architectures. This poster will report the formulations, structures, and preparation conditions that yield the best orientations and properties that are achievable using these embodiments.

In-Situ TEM Experiment of Ferroelectric Switching in AlBN

There is growing research interest in doped wurtzite-structured thin films that exhibit ferroelectricity, with the hope of integrating ferroelectricity into complementary metal-oxide semiconductors (CMOS) processes. Among these materials, B-substituted AlN thin films have shown great potential as integrated ferroelectric materials. This ferroelectric nitride maintains the polar wurtzite structure of AlN, has a relatively high bandgap, and exhibits polarization values larger than 125 μC cm-2. An ongoing research challenge is to lower the coercive field below the ~5MV/cm benchmark value. This work strives to understand the role/mechanism by which Boron influences the switching mechanisms and coercive field via in-situ transmission electron microscopy (TEM). A major challenge in the experimental approach is the need to apply very high electric fields to match the coercive field strength, which is close to the break-down strength of the material. In addition, such high electric fields can lead to surface electromigration of the electrodes, and thus shorting, of the TEM specimens. In this work, we explore different sample preparation techniques and geometries that enable in-situ switching experiments on these technologically important ferroelectric thin-films.

Magnetism in thin films of the high entropy oxide La(Cr0.2Mn0.2Fe0.2Co0.2Ni0.2)O3

Information technologies account for ~8% of global energy use and are projected to account for up to 14% of global greenhouse gas emissions by 2040. Operating on electron spin instead of charge, spintronic devices offer fast, non-volatile, and energy efficient computing and memory. La(Cr0.2Mn0.2Fe0.2Co0.2Ni0.2)O3 (L5BO) is a high entropy perovskite oxide which has attracted interest due to the ability to switch between antiferromagnetic (AFM) and ferromagnetic (FM) states by varying its Mn concentration. By targeting the boundary between these two states, it may be possible to use other stimuli to switch between AFM and FM, allowing for new spintronic functionality. This switching behaviour is due to competition between the 15 different exchange interactions in this material. By growing epitaxial thin films of L5BO, we aim to investigate how the interplay between competing interactions can be used to engineer magnetic properties through thin film heterostructures and finely controlling stoichiometry, thickness, and strain. We have synthesized this material with varying concentrations of Mn and suggest that this leads to a mixed AFM and FM based on preliminary powder magnetometry results. We also demonstrate that high-quality thin films of this material grow epitaxially on SrTiO3, LaAlO3, and NdGaO3 substrates based on atomic force microscopy and X-ray diffraction measurements.

Critical Template Characteristics for Efficient and Reproducible Templated Grain Growth (TGG) of Textured Piezoelectric Ceramics

Recently textured piezoelectric ceramics have attracted attention since they demonstrate better mechanical properties and comparable piezoelectrical with respect to single crystals. Textured ceramics can be manufactured via TGG process. In this process, template particles (anisometric large seeds), distributed among the fine equiaxed matrix grains, are aligned during the shaping process. During heat treatment, the templates grow at the expense of fine matrix particles via Ostwald ripening process. Since they maintain their initial alignment, the final microstructure is composed of oriented grains and hence exhibits textured character. In the TGG process, the templates are the key components to induce texture. Template characteristics can play critical role on the properties of the textured ceramics. Therefore, it is important to develop an understanding on the relationship between template properties and texture development. Accordingly, in this presentation, effects of barium titanate, sodium niobate and sodium bismuth titanate template characteristics (i.e., size, shape, chemistry, and degree of lattice matching with the matrix grains) on texture development and piezoelectrical properties in piezoelectric ceramics will be discussed. The critical template and processing parameters for the efficient and reproducible TGG will be also elaborated.

AUTHOR INDEX

Abbott, T. 36
Acharya, M. 96
Adabifiroozjaei, E. 111
Adamo, C. 27
Ahadi, K. 43, 73
Ahn, C. 35
Ahn, J. 127, 137
Ahn, Y. 27
Alem, N. 50
Alema, F. 57
Alexe, M. 86
Alff, L. 105, 111
Almazan, I. 38
Almishal, S. S. 44, 50
Al-Tawhid, A. 43, 73
Ameres, M. 7
An, K. 113
Anderson, K. P. 16
Andrew, J. 16, 30
Andrews, J. 5
Anguish, A. 16
Anh, J. 125, 126
Arcon, D. 69
Arh, T. 69
Arndt, N. 33
Arnold, D. 30
Aryana, K. 60
Arzumanov, A. 111
Aumen, A. 94
Ayyagari, S. V. 50
Babcock, S. E. 36
Bachman, W. 91
Bagués, N. 29
Bahoura, M. 121
Baker, T. 8
Bansal, M. 29
Bansil, A. 70
Barber, J. P. 51
Barbiellini, B. 70
Barnard, J. P. 28
Bas, D. 80
Basit, N. 36
Bates, S. 116
Beall, R. 140
Bejger, G. R. 46
Bellcase, L. 2
Bishop, S. 8, 91
Blea-Kirby, M. 91
Bock, J. A. 91
Boschker, H. 41
Boston, R. 5
Bowden, M. 37, 40
Brahlek, M. 68, 109
Braun, M. 63
Broughton, R. 15, 140
Brown, M. 58
Bruder, E. 86
Bryant, G. 124
Buckner, H. 85
Bullard, T. 67
Bulmer, J. 67
Burns, S. 131
Burzynski, K. 58
Butala, M. 33
Butun, S. 36
Buzanich, A. 115
Cai, Z. 36
Calderon, S. 142
Cann, D. 133
Canton, J. 95
Cao, Y. 107
Carnis, J. 27
Carr, C. 107
Casamento, J. 141
Caucci, M. 46
Cha, W. 107
Chae, S. 4, 49
Chambers, S. 37, 40
Chen, A. 61, 99, 100
Chen, Y. 113
Chilcote, M. 59, 68
Cho, J. 76
Choi, J. 125
Choi, S. 3
Choudhary, K. 66
Chow, P. 136
Chun, S. 27
Clem, P. 132
Colfer, L. 29
Collazo, R. 55
Corbett, J. P. 58
Culbertson, C. M. 8
Dabo, I. 50
Dai, Z. 11
Danquah, S. 121
De Souza, R. A. 81

Dean, A. 8
Degen, C. 22
Denize, C. 121
Dickey, E. C. 94, 122, 142
Dilley, N. R. 28
Dionne, C. J. 134
Dolgos, M. 131
Dolgos, T. Rowem. 128
Donahue, P. 107
Dong, Y. 38
Dou, H. 28
Du, Y. 37, 40, 84
Dubey, A. 115
Dufresne, E. 38
Dukes, M. Dressel 117
Dumitrescu, E. 72
Eom, I. 27
Escobar, M. 115
Evans, A. M. 134
Evans, P. G. 27, 36
Eyink, K. 58
Feigelson, B. N. 16
Ferson, N. D. 30
Feygelson, T. I. 52
Fields, S. 14
Foley, B. 134
Fong, D. 38
Fortier, E. C. 141
Frame, M. 94
Franson, A. 80
Frazer, T. D. 107
Frömling, T. 86, 87
Frye, M. 98
Fulanovic, L. 7
Gabilondo, E. 15
Gajjala, S. 17, 18
Ganesh, P. 38
Ganiban, J. 30
Garten, L. M. 98
Gilbert, D. A. 103
Giri, A. 134
Go, S. 4
Gonzalez, A. 43, 73
Goodling, D. K. 21, 141
Gopalan, V. 50, 76
Gosar, Z. 69
Grobis, M. 60
Guan, X. 123
Guerrero-Sanchez, J. 58
Guo, Y. 117
Gupta, J. Jay 58
Gyan, D. Sri 27, 36
Hagemann, U. 115
Haile, S. M. 107
Halasz, G. 72
Harmon, K. 107
Harper, C. J. 2
Hatton, P. 84
Haugan, T. J. 67
Hayden, J. 13, 141, 142
He, J. 76
He, Z. 28
Hemley, R. 136
Heron, J. 44, 49, 50, 51
Hinterstein, M. 63
Hirt, D. 60
Ho, D. 124
Hofbeck, I. 133
Hoffmann, M. J. 63
Höfling, M. 90
Holt, M. 36
Hong, S. 122
Hopkins, P. 60, 134
Hossain, M. 40
Hruszkewycz, S. 107
Hu, G. 38
Hu, R. 124
Hu, X. 36
Hua, C. 72
Huang, X. 96
Hubbard, W. 111
Huey, B. 61
Huxter, W. 22
Ihlefeld, J. 14, 61
Ikuhara, Y. 88
Islam, M. 60
Iyer, A. 76
Jang, J. 76
Janotti, A. 124
Jaszewski, S. T. 14, 61
Jauregui, L. 91
Jeen, H. 38, 39
Jena, D. 45
Jeon, I. 62
Jeong, M. 127, 137
Jia, Q. 100
Jiang, T. 111
Jo, J. 27
Jo, W. 3
Jones, A. H. 134
Jones, J. L. 2, 15, 140
Kabat, N. 36
Kaiser, N. 105
Kalaswad, M. 28

Kanatzidis, M. 76
Kaspar, T. 37, 84
Keat, C. 115
Keeney, L. 29
Keppens, V. 33
Kezer, P. 44, 50
Khan, S. A. 3
Kholkin, A. 23
Kim, D. 125
Kim, E. 4, 6
Kim, H. 4, 126
Kim, J. 34
Kim, S. 27
Kim, T. 27
Kioupakis, E. 49
Ko, W. 72
Koc, R. 17, 18
Koirala, K. P. 37, 40
Komissinsky, P. 111
Kotsonis, G. 46
Kovalenko, S. 115
Krause, A. 119
Krogel, J. 38
Kuech, T. F. 36
Kumar, R. 136
Laing, R. 58
Lamas, R. Rodriguez 90
Landahl, E. 27
Lane, C. 70
Lang, A 52
Langhout, J. 119
Laskin, G. 41
Lawrie, B. 72
Lee, H. 27, 38, 42
Lee, J. 3, 27, 49, 125
Lee, S. 3, 127, 137
Leighton, C. 101
Lenox, M. K. 14
Lewin, D. 130
Li, J. 83, 139
Li, N. 27
Li, P. 79
Li, Y. 83
Lin, G. 77
Liu, H. 38
Liu, J. 32
Liu, R. 36
Liu, W. 78
Liu, Y. 107
Liu, Z. 77
Long, B. 29
Lord, M. T. 143
Lu, S. 36
Lupascu, D. C. 115, 130
Lv, B. 69, 78
Ma, J. 77
Maclaughlin, E. 8
Maddali, S. 107
Maekawa, T. 56
Maggard, P. 15
Mahalingam, K. 58
Mahmoudvand, M. 131
Maity, T. 29
Maksymovych, P. 71, 72
Mannhart, J. 41
Maria, J. 13, 21, 44, 46, 47, 50, 55, 108, 141, 142
Marimon, N. Domingo 25
Markiewicz, R. S. 70
Marks, S. 36
Martin, L. W. 96
Martinez, R. 132
Marusak, K. 117
Maximenko, S. 52
Mazza, A. 33, 64, 112
McComb, D. W. 29
McConnell, J. 117
McGuire, M. 68
Meirzadeh, E. 134
Menze, K. 130
Meyer, D. J. 52
Miao, H. 68
Michiels, I. 130
Milich, M. 134
Mills, S. C. 16, 98
Mischenko, D. A. 51
Mischenko, D. 44
Moering, J. 110
Mohammed, A. Z. 131
Molina-Luna, L. 105, 111
Moore, R. 68
Muhammad, Q. 87
Muralikrishnan, V. 119
Muratore, T. 58
Musicó, B. L. 33
Nag, J. 60
Nahm, S. 4, 6
Nasiou, D. 105, 111
Nathan, S. 120
Need, R. 33
Nepal, N. 52
Neuman, E. 8
Newburger, M. 80
Newell, R. 15
Niculescu, G. 51

Nordlander, J. 55
Nuckolls, C. 134
O'Donnell, S. 15
Oh, S. 68
Oh, Y. 126
Okamoto, S. 68
Osinsky, A. 57
Osofsky, M. 79
Oxley, B. 76
Padgett, A. S. 129, 132
Padula, A. P. Gayoso 90
Page, M. 80
Pai, Y. 68
Paik, H. 45
Paldi, R. L. 28
Pan, H. 96
Park, H. 3
Park, S. 27
Pate, B. B. 52
Patterson, E. 16
Paudel, B. 37
Payne, A. 140
Peretti, A. 91
Perez-Garza, H. 111
Perry, N. H. 85, 89
Piros, E. 105
Pivak, Y. 111
Poage, S. 43, 73
Popescu, A. 107
Porz, L. 86
Pradhan, S. 121
Pramudya, Y. 63
Prestigiacomo, J. 79
Price, P. M. 8
Puri, K. 33
Ræder, T. 90
Ram, A. 124
Ramasubramanian, L. Narayanan 54
Randall, C. 92, 93
Rao, B. A. 31
Rappe, A. M. 11, 20
Reboredo, F. 38
Recalde-Benitez, O. 105, 111
Reed, A. 58
Regan, B. 111
Richtik, B. N. 128
Rocha, J. 23
Rödel, J. 86, 87
Romanyuk, K. 23
Rondinelli, J. 76, 120
Ross, C. 102
Rost, C. M. 44, 46, 51
Roy, C. 8
Roy, P. 100
Roy, X. 134
Ruan, Y. 111
Sahebkar, K. 33
Saito, K. 116
Samarakoon, W. 40
Sando, D. 24, 143
Sandy, A. 38
Saritas, K. 38
Sarma, R. 28
Sarott, F. 22
Sarott, M. F. 19
Savage, D. E. 36
Scherer, M. 7
Schlom, D. 27, 45
Schmidt, M. 29
Schreiber, D. 84
Sebastian, M. 67
Shah, P. 80
Sheehan, B. 29
Shen, J. 28
Shi, J. 53
Shukla, N. 14
Shur, V. 23
Shvartsman, V. V. 115
Siddiqui, A. 28
Simons, H. 90
Simpson-Gomez, J. 85
Sinclair, D. C. 95
Singh, M. 54
Singh, R. N. 54
Sinnott, S. 46
Sitar, Z. 55
Skidmore, C. 108
Smith, C. S. 97
Soleimany, M. 86
Son, Y. 48
Song, J. 61
Song, S. 72
Spurgeon, S. R. 84
Spurling, R. 47
Sreenivas, V. 58
Srot, V. 41
Stähler, J. 115
Starchenko, V. 38
Staruch, M. 98
Stewart, D. 60
Storm, D. F. 52
Strimaitis, J. 121
Stueckler, M. 118
Sun, J. 70

Sun, W. 83
Sushko, P. 37, 40
Suvaci, E. 144
Suzuki, Y. 79
Swift, G. 17, 18
Tang, Q. 9, 10
Taylor, S. 84
Tian, Z. 96
Tilles, J. N. 132
Tomko, J. 60
Tonks, M. R. 123
Trassin, M. 22, 104
Trenary, M. 136
Trolier-McKinstry, S. 13, 48
Tzou, W. 142
Uberuaga, B. P. 84
Unithrattil, S. 27
Upadhyay, S. 12
Ushakov, A. 23
Valanoor, N. 143
Van Aken, P. 41
Van Omme, T. 111
Verma, A. 45
Vieira, F. M. 50
Vitale, B. J. 8
Vogel, T. 105
Wagner, S. 63
Walden, S. 117
Wan, G. 38
Wang, C. 138, 139
Wang, H. 28, 41
Wang, L. 37, 40, 82, 84
Wang, Q. 138, 139
Wang, S. 61
Wang, Y. 114
Ward, T. Z. 33
Waters, M. 76
Webb, M. 51
Weber, M. C. 26
Williams, L. 49
Winkler, R. 105, 111
Winner, D. 58
Wolf, M. 80
Wollmershauser, J. 16
Wu, W. 74
Xiao, Y. 136
Xie, B. 135
Xie, S. 65
Xu, W. 23
Yan, J. 72
Yang, W. 37, 40
Yano, K. 84
Yao, X. 68
Ye, P. 28
Ye, Z. 114
You, L. 61
Yousefian, P. 92, 93
Yusenko, K. 115
Zakutayev, A. 1
Zelenovskiy, P. 23
Zeng, H. 100
Zhang, J. 20
Zhang, Q. 33, 38
Zhang, R. 70
Zhang, Y. 28
Zhao, X. 138
Zheng, D. 28
Zhou, H. 38, 40, 75, 77, 85
Zhou, S. 20
Zhou, T. 36
Zhu, J. 70
Zhu, W. 48
Zhu, X. 9
Zhu, Y. 106
Zhuo, Z. 37
Zintler, A. 105, 111
Zorko, A. 69
Zschiesche, D. 8
Zu, R. 50
Zuo, P. 36